MOS Interface Physics, Process and Characterization

MOS Interface Physics, Process and Characterization

Shengkai Wang and
Xiaolei Wang

CRC Press
Taylor & Francis Group
Boca Raton London New York

CRC Press is an imprint of the
Taylor & Francis Group, an **informa** business

First edition published 2022
by CRC Press
6000 Broken Sound Parkway NW, Suite 300, Boca Raton, FL 33487-2742

and by CRC Press
2 Park Square, Milton Park, Abingdon, Oxon, OX14 4RN

Library of Congress Cataloging-in-Publication Data
Names: Wang, Shengkai, 1984– author. | Wang, Xiaolei, 1985– author.
Title: MOS interface physics, process and characterization / Shengkai Wang, Xiaolei Wang.
Description: First edition. | Boca Raton CRC Press, 2022. | Includes bibliographical references. |
Summary: "The electronic device based on Metal Oxide Semiconductor (MOS) structure is the most important component of a large-scale integrated circuit and the key to achieving high performance devices and integrated circuits is high quality MOS structure. This book contains abundant experimental examples focusing on MOS structure.
The volume will be an essential reference for academics and postgraduates within the field of microelectronics"— Provided by publisher.
Identifiers: LCCN 2021017023 (print) | LCCN 2021017024 (ebook) |
ISBN 9781032106274 (hbk) | ISBN 9781032106281 (pbk) | ISBN 9781003216285 (ebk)
Subjects: LCSH: Metal oxide semiconductors—Design and construction—Mathematics. |
Semiconductors—Junctions. | Integrated circuits—Research. | Solid state physics—Experiments.
Classification: LCC TK7871.99.M44 W358 2022 (print) |
LCC TK7871.99.M44 (ebook) | DDC 621.3815/284—dc23
LC record available at https://lccn.loc.gov/2021017023
LC ebook record available at https://lccn.loc.gov/2021017024

ISBN: 978-1-032-10627-4 (hbk)
ISBN: 978-1-032-10628-1 (pbk)
ISBN: 978-1-003-21628-5 (ebk)

DOI: 10.1201/9781003216285

Typeset in Minion
by codeMantra

Contents

Preface

B OTH OF THE AUTHORS of this book are engaged in metal–oxide–semi-conductor (MOS) interface research. We have always wanted to write a short, clear and easy-to-use book to help students answer their doubts in practice and carry out relevant work as soon as possible. In the long-term research and guidance of students, we found such a common problem; that is, although there are many books on physics or semiconductors on the market, students are often faced with difficulties in getting these books. They do not have a thorough understanding of MOS interface physics, they are not familiar with the preparation skills and key technologies of devices, and more importantly, they lack knowledge because there are few practical examples that can be referred to, so that sometimes it is difficult to prepare high-quality MOS device structure, sometimes it is not clear how to debug the relevant equipment, and even it is difficult to determine whether the data are accurate. This brings some difficulties for our teaching and students to enter the scientific research.

Here is a simple example. Students often encounter such problems: In the preparation process of MOS devices, the contact resistance of the back electrode puzzles everyone. How to prepare high-quality back contact by simple methods and how to calibrate the equipment to remove the parasitic resistance have become very important problems, but this is often not involved in the common textbooks.

Based on our experience at the Institute of Microelectronics of the Chinese Academy of Sciences, and combined with some problems accumulated in the early stage, we condensed what we wanted to express to the readers into this short book. We hope to introduce the MOS interface physics, process and characterization to readers through simple expression and try not to use profound formulas and obscure words.

In the process of writing this book, we have received a lot of support by many people. This book would not have been completed without them. First of all, we would like to thank the staff from the publisher Taylor & Francis Group, especially Ms. SUN Lian and Ms. CHEN Jingying, and Ms. Vaishnavi Venkatesan from codeMantra for processing our book. In addition, in the process of writing the book, our colleagues and students also provided a lot of advices and examples. They are Mr. YAO Peilin, Dr. HAO Jilong, Ms. YOU Nannan, Ms. ZHANG Qian, Mr. LIU Peng, Mr. HU Qingyu, Ms. CAO Qianqian, and Mr. Zhang Ziqi. We would also like to thank the Youth Innovation Promotion Association of the Chinese Academy of Sciences for its support for this book.

WANG Sheng-Kai & WANG Xiao-Lei
Feb. 2021 in Beijing

Authors

Shengkai Wang is a professor in the Institute of Microelectronics, Chinese Academy of Sciences. He received Ph.D. from the University of Tokyo in 2011 and has been engaged in Ge, III-V, SiC MOS technology. He has published more than 100 papers and authorized 40+ patents.

Xiaolei Wang is a professor in the Institute of Microelectronics, Chinese Academy of Sciences. He received Ph.D. from the Institute of Microelectronics, Chinese Academy of Sciences in 2013 and has been engaged in Si/Ge based MOS technology. He has published more than 100 papers.

Introduction

0.1 SCOPE AND PLAN OF THE BOOK

As the most important and widely used field-effect structure, metal–oxide–semiconductor (MOS) is the core component of modern integrated circuits. For the understanding of MOS system, this book aims at introducing the MOS interface physics, process and characterization to readers practically through simple expressions without profound formulas and obscure words. As mentioned in the book title, it mainly covers the related contents from three aspects: interface physics of MOS, processes of MOS and characterizations of MOS. These three parts constitute the main chapters of the book, Chapters 1–3.

Chapter 1 introduces the interface physics of MOS devices, including what is a MOS interface, the physical nature of interface states and bulk defects, the passivation idea of MOS interface, interface mechanics, thermodynamics, material chemistry and other physical phenomena in MOS devices.

Chapter 2 introduces the process technologies of MOS devices, including how to prepare a high-quality MOS with detailed experimental skills, thermal oxidation process and model of Si and other materials, mechanism of typical deposition methods and equipment in MOS devices.

Chapter 3 introduces the characterizations of MOS devices, including methods for evaluating the density of interface states of MOS, experimental skills in MOS characterization, hysteresis and bulk charge in MOS, how to extract equivalent oxide thickness of MOS, what gate leakage stands for and how to measure it, and the characterization of work function.

DOI: 10.1201/9781003216285-1

Since this book aims at the practical dimension, physical formulas and theoretical content of characterization have not been included. For such contents, some classic books, for example, "Physics of Semiconductor Devices" by Prof. S. M. Sze and Kowk. K. Ng or "MOS Physics and Technology" by E. H. Nicollian and J. R. Brews, could be referred to.

0.2 BRIEF HISTORY OF MOS DEVICES

The invention and application of integrated circuit is the most brilliant pearl in the history of science and technology in the 20th century. Over the past 60 years, integrated circuits have not only brought great success to economic prosperity, social progress and national security, but also changed people's production, life and way of thinking. At present, integrated circuits exist everywhere and all the time. She has become an indispensable part of human civilization.

For modern integrated circuits, a MOS device is a very magical part. From the philosophical point of view, it is the embodiment of human wisdom and contains the full and harmonious use of nature. As we all know, the top three elements in our earth are oxygen, silicon and aluminum. The early integrated circuit is made by the repeated use of these three elements, in which silicon constitutes the semiconductor, silicon oxide constitutes the oxide layer and aluminum acts as the metal layer and is also used as the interconnection material.

The exponential progress in MOS technology, called Moore's law, is illustrated by the evolution of the number of MOS transistors integrated on a single memory chip or a single microprocessor, measured in calendar years. The increase in integration density is mainly due to the scaling-down of transistor size. Recently, TSMC officially disclosed the latest details of 3-nm process technology, with transistor density as high as 250 million per square millimeter. Of course, the proposal of this great structure has gone through a long incubation period, which can be traced back to the early 20th century. So first of all, let us take a look at the development history of MOS structure.

Due to technical limitations, a device called the IGFET was first described in patents by Lilienfeld and Heil in the 1930s. Julius Edgar Lilienfeld was born in Lemberg, Austria-Hungary. He proposed a device similar to the modern metal semiconductor field-effect transistor (MESFET) in 1926. A few years later, he proposed a device similar to the modern MOS transistor in 1928. In 1932, Igor Yevgenyevich Tamm proposed a concept known as surface states, which is one of the central topics of this book. Surface states according to Tamm's theory are known as Tamm states. Later in 1935, Oskar Heil, a German scientist, proposed the idea of controlling the resistance in a semiconducting material with an electric field in British patent 439,457. Note that the pioneering patents by Lilienfeld and Heil belong to "idea-patent", and they were not practical due to the technology limitations. Later in 1939, Shockley proposed his theory on surface states. Note that surface states that are calculated in the framework of a tight-binding model are often called Tamm states, while surface states that are calculated in the framework of the nearly free electron approximation are called Shockley states. In contrast to the nearly free electron model used to describe the Shockley states, the Tamm states are suitable to describe also transition metals and wide-gap semiconductors. On December 23, 1947, Bardeen and Brattain invented the point-contact transistor, which became a great milestone in the history of mankind. In 1952, Shockley published a theoretical paper, indicating the debut of field-effect transistors in the form of a junction field-effect transistor (JFET).

In 1959, Martin M. Atalla made the first practical MOS transistor based on Si, after systematical investigation on the surface passivation of silicon surface by SiO_2. Later, he assigned the task to Dawon Kahng, a scientist in his group. Eventually, Attalla and Kahng announced their successful MOSFET at a 1960 conference. In 1963, Frank Marion Wanlass from Fairchild Semiconductor invented the first CMOS under US patent 3,345,858.

In 1965, many achievements have been made. B. E. Deal and A. S. Grove published their famous paper on "General Relationship for the Thermal Oxidation of Silicon" that pushed the application of SiO_2/Si forward. Besides the use of the SiO_2/Si system, P. Balk reported that hydrogen annealing was another important technical development to lower the density of interface states. Also in this year, Gordon Moore made his famous predication "Moore's law". In 1967, R. E. Kerwin et al. first put the Si gate technology for ICs under US patent 3,475,234, and in 1974, R. H. Dennard proposed the quantified scaling rule of IC process design, called Dennard's rule. Guided by Moore's law and Dennard's rule, intentionally or not, the integrated circuit industry took off to incredible proportions and has become one of the world's leading industries. The milestone events in the history of MOS devices are summarized in Table 0.1.

TABLE 0.1   Milestone Events in the History of MOS Devices

Year	People	Event	Reference
1926	J. E. Lilienfeld	The first patent of a device similar to the modern MESFET	US patent 1,745,175, filed in 1926 and awarded in 1930
1928	J. E. Lilienfeld	The first patent of a device similar to the modern MOS transistor	US patent 1,900,018, filed in 1928 and awarded in 1933
1932	I. Tamm	Proposal of a concept known as surface states	Phys. Z. Soviet Union, vol. 1 (1932), pp. 733–746
1935	O. Heil	A device controlling the resistance in a semiconducting material with an electric field	British patent 439,457, filed in 1935 and awarded in 1935
1939	W. Shockley	On the surface states associated with a periodic potential	Phys. Rev., vol. 56, 317-323 (1939)
1947	J. Bardeen and W. Brattain	Invention of point-contact transistors	Bell Labs logbook (December 1947), pp. 7-8, 24
1953	W. Shockley	Invention of junction field-effect transistors (JFETs)	Proc. IRE, vol. 40, no. 11 (November 1952), pp. 1365–1376.
1959	M. M. Atalla	Passivation of silicon surface state by SiO_2	Bell System Technical Journal, vol. 38, no. 3 (May 1959), pp. 749–783
1960	D. Kahng and M. M. Atalla	The first practical MOS transistor based on Si	IRE/AIEE Solid-State Device Research Conference, USA, 1960
1963	F. Wanlass	Invention of the first CMOS	US patent 3,356,858, filed in 1963 and awarded in 1967
1965	B. E. Deal and A. S. Grove	General Relationship for the Thermal Oxidation of Silicon	J. Appl. Phys. 36, 3770 (1965)
1965	G. Moore	Moore's law	Electronics Magazine vol. 38, no. 8 (April 19, 1965).
1965	P. Balk	Hydrogen annealing to lower the density of surface states	Electrochemical Society Spring Meeting, San Francisco, California, USA, 1965
1967	R. E. Kerwin et al.	Silicon gate technology developed for ICs	US patent 3475234 (filed March 27, 1967, and issued October 28, 1969)
1974	R. H. Dennard et al.	Scaling of IC process design rules quantified (Dennard's rule)	IEEE Journal of Solid-State Circuits, Vol. 9 (October 1974), pp. 256-268

BIBLIOGRAPHY

1. Lilienfeld, J.E., Method and apparatus for controlling electric currents, US Patent 1,745,175, filed in 1926 and awarded in 1930.
2. Lilienfeld, J.E., Device for controlling electric currents, US Patent 1,900,018, filed in 1928 and awarded in 1933.
3. Heil, O., Improvements in or relating to electrical amplifiers and other control arrangements and devices, British Patent 439,457, filed in 1935 and awarded in 1935.
4. Riordan, M., and L. Hoddeson, *Crystal Fire: The Invention of the Transistor and the Birth of the Information Age*, (New York, Norton, 1997).
5. Tamm, I., Uber eine mogliche art der elektronenbindung an kristallober-flachen. *Physikalische Zeitschrift der Sowjetunion*, 1932. **1**, p. 733–746. (Note: The title of the paper after translation from German to English is "On the possible bound states of electrons on a crystal surface". The full name of the journal is Physik Zeitschrift der owjetunion. This paper was written in German and has been re-printed in "I.E. Tamm Selected Works", edited by B.M. Bolotovskii and V. Ya. Frenkel, Springer-Verlag, Berlin 1991, pp. 92–102.)
6. Shockley, W., On the surface states associated with a periodic potential. *Physical Review*, 1939 **56**(4): p. 317–323.
7. Shockley, W., A unipolar field-effect transistor. *Proceedings of the IEEE*, 1952, Nov. **40**(11): p. 1365–1376.
8. Dacey, G.C., and I. M. Ross, Unipolar field-effect transistor. *Proceedings of the IEEE*, 1953, Aug. **41**(8): p. 970–979.
9. Mead, C. A., Schottky barrier gate field effect transistor. *Proceedings of the IEEE*, 1966, Feb. **54**(2): p. 307–309.
10. Atalla, M.M., E. Tannenbaum, and E.J. Scheibner, Stabilization of silicon surfaces by thermally grown oxides. *Bell System Technical Journal*, 1959, May. **38**(3): p. 749–783.
11. Kahng, D., and M.M. Atalla, Silicon-silicon dioxide field induced surface devices. *IRE/AIEE Solid-State Device Research Conference*, Carnegie Institute of Technology, Pittsburgh, PA, 1960.
12. Kahng, D., Silicon-silicon dioxide surface device, Technical memorandum of Bell Laboratories issued on January 16, 1961. This paper has been reprinted in the book Semiconductor Devices: Pioneering Papers, edited by S. M. Sze, World Scientific, Singapore, 1991: D. Kahng, "Silicon-silicon dioxide surface device", p. 583–596.
13. Kahng, D., Electric controlled semiconductor device, US Patent 3,102,230, filed in 1960 and awarded in 1963.
14. Kahng, D., A historical perspective on the development of MOS transistors and related devices. *IEEE Transactions on Electron Devices*, 1976, Jul. **23**(7): p. 655–657.
15. Balk, P., *Effects of Hydrogen Annealing on Silicon Surfaces*, (Electrochemical Society Spring Meeting, San Francisco, CA, 1965).

16. Nicollian, E.H. and J.R. Brews, *MOS Physics and Technology*, (Wiley, New York, 2003), p. 1–906.
17. Riezenman, M.J Wanlass's CMOS circuit. *IEEE Spectrum*, 1991, May (**28**(5), p. 44.

CHAPTER 1

Physics of Interface

1.1 MOS INTERFACE

The metal–oxide–semiconductor (MOS) interface refers to the hetero-junction interface in the gate structure of the MOS device, as shown in Figure 1.1. The gate structure of the MOS device is usually composed of a metal/oxide dielectric/semiconductor substrate, and the oxide dielectric can be a stack of one or several insulating dielectrics. The interface appears between different kinds of materials. Therefore, the interface can be considered as the boundary between any two different materials. It should be noted that in most cases, the interface does not refer to a surface of infinitesimal thickness, but refers to the transition area between two different materials, and it is a thin layer with a certain thickness, usually about 3 Å.

The following types of interfaces in the MOS gate structure can appear: oxide/semiconductor interface, oxide/oxide interface and metal/oxide interface. Generally speaking, the characteristics of the oxide/

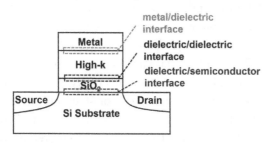

FIGURE 1.1   Schematic of modern metal–oxide–semiconductor (MOS) device.

DOI: 10.1201/9781003216285-2

semiconductor interface are the most important. It seriously affects the gate control capability of the MOS devices, the mobility of carriers on the semiconductor substrate and the reliability of the MOS gate structure. In addition, after the introduction of high dielectric constant gate dielectric (such as HfO_2) into the gate structure of silicon-based MOS devices, the HfO_2/SiO_2 interface also plays an important role, involving the shift of the device threshold voltage and the reliability of the gate structure.

1.2 THE PHYSICAL NATURE OF INTERFACE STATES AND BULK DEFECTS

The concept of interface states does not belong to the category of classical physics, but belong to the category of quantum mechanics and solid-state physics. The interface state refers to the real-space distribution of the electronic wave function near the interface, i.e., the electronic wave function attenuates to both sides of the interface. The electronic energy level corresponding to the interface state is usually located inside the band gap. The study of the interface states has been going on for nearly hundred years, but it is still not fully understood. The research on the interface state of silicon semiconductor is the most complete, and the understanding is the most profound. Here we take the interface state of silicon as an example. The physical origin of the interface state of silicon is often attributed to dangling bonds, or P_b centers. This concept is actually a visual explanation given from a chemical point of view. From a physical point of view, it needs to be considered from the energy band point of view. The interface state is actually not generated out of thin air, but the energy level in the conduction band or valence band of silicon is pulled into the forbidden band, and then becomes the interface state, as shown in Figure 1.2.

Due to the interrupt of periodicity at the SiO_2/Si interface, the solutions of Schrödinger's equation with complex wave vectors become of physical relevance for energies within band gaps, resulting in gap states at the SiO_2/Si interface. These interfacial gap states are derived from the virtual gap states of the complex band structure of the silicon semiconductor, and they may arise from intrinsic, defect, or structure induced gap states. They consist of valence- and conduction-band states. The characteristics of these gap states change across the band gap from predominately donor- to acceptor-like closer to the valence band top and the conduction band bottom, respectively. The energy at which their characteristic changes is called their branch point, or most generally, charge neutrality level (CNL).

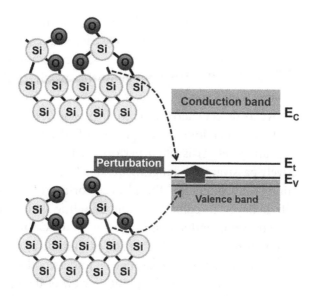

FIGURE 1.2   Origin of interface trap.

This energy level shift comes from the change of the potential energy at the interface relative to the potential energy inside the silicon. The physical sources of these changes include dangling bonds and interface structure relaxation. It should be noted here that certain chemical bonds can also lead to the interface state energy level. For example, the Ga-O bond on the surface of InGaAS has the bond energy in the InGaAS forbidden band. Although there is no dangling bond, the bond energy position deviates from the conduction. Band or valence band can still lead to interface states.

The physical nature of bulk defects originates from atomic vacancies, interstitial atoms, replacement atoms, dislocations, structural changes, etc. Any deviation from the perfect structure of the material may cause defects. Corresponding to the gate structure of MOS devices, body defects often involve oxygen vacancies. Similar to the generation process of the interface state, the body defect energy level also pulls the electron energy level from the conduction or valence band into the forbidden band.

1.3 MOS INTERFACE PASSIVATION METHODS

The method of interface passivation comes from the physical nature of interface defects. At present, the most well-researched semiconductor substrates include silicon, germanium, silicon germanium and III-IV semiconductors such as InGaAS. The origins of the interface states of these semiconductors are slightly different, so they need to be treated separately. The following describes the passivation ideas of each semiconductor.

For silicon semiconductor, dangling bonds are the main source of the interface states. Therefore, the dangling bonds that passivate silicon are the guiding ideology. Hydrogen atoms can passivate the dangling bonds of silicon and move its defect energy level to the inside of the conduction band or valence band. Therefore, in the process of silicon MOS devices, forming gas annealing (FGA) is often performed in a hydrogen atmosphere to effectively passivate the dangling bonds of silicon.

For germanium semiconductor, the physical origin of the interface state is generally believed to come from the dangling bonds of germanium. Unlike silicon semiconductors, hydrogen atoms cannot passivate the germanium dangling bonds well. This shows that the physical origin and passivation of the interface state cannot be understood simply from a chemical point of view, i.e., from the point of view of dangling bonds or atomic bonding. It still needs to return to deeper physical methods, such as first-principles calculations, to accurately understand the physical origin and passivation of interface states. This inconvenience is the lack of easy-to-understand physical images to guide actual projects, especially for engineers and technicians, which is difficult to understand.

For silicon germanium semiconductor, the physical origin of the interface state is mainly related to the Ge-O bond at the interface. The fewer Ge-O bonds, the smaller the interface state. This shows that the Ge-O bond can induce interface states, although it does not produce the dangling bonds. Therefore, inhibiting the formation of Ge-O bonds is the guiding ideology of silicon germanium semiconductor passivation. This involves the interface thermodynamic/thermodynamic process, i.e., the oxidation process of the silicon germanium semiconductor. Current research has found that silicon oxide has a smaller Gibbs free energy than germanium oxide, which means it is easier to form. Therefore, annealing at an appropriate temperature (such as about 500°) can transform the oxide of germanium to the oxide of silicon, which is beneficial to reduce the interface state density. Another passivation idea is to prevent the formation of Ge-O bonds, such as epitaxial silicon thin layers on silicon germanium. This method can achieve the interface state density comparable to silicon passivation levels.

For III-IV semiconductors such as InGaAS, the physical origin of the interface state is the existence of Ga-O bonds. Therefore, inhibiting the formation of Ga-O bond is the guiding method for passivation. This also involves the interface thermodynamics/thermodynamic processes; however, the research in this area is still not clear.

1.4 INTERFACE THERMODYNAMICS

The interface thermodynamics/thermodynamic process of semiconductors is still not fully understood and mastered. The research on the interface thermodynamics of silicon is the most profound. Therefore, here is an example of silicon to introduce the research progress of its interface thermodynamics. During the thermal oxidation and growth of SiO_2 on the silicon substrate, a transition layer appears between the silicon substrate and SiO_2. The atomic ratio of oxygen to silicon in this transition layer is less than 2, and the space thickness is about 7 Å. Dissociative adsorption of O_2 molecule occurs via a charge transfer at the dangling bond site not only on Si surfaces but also at SiO_2/Si interfaces. During the oxidation, the chemically active dangling bond is persistently supplied at the SiO_2/Si interface by the pint defect generation (emitted Si atom and vacancy) due to the intrinsic (oxidation-induced) and extrinsic (thermally induced) strain with assistance of the heat of adsorption and the thermal activation.

The high dielectric constant gate dielectric has been widely doped into the MOSFET devices, and the interface between the high dielectric constant gate dielectric and the silicon substrate is discussed here. Here we take hafnium oxide as an example for discussion. In experiments, there appears SiO_x or HfSiO between the HfO_2 and silicon substrates, which indicates that HfO_2 and Si are prone to reaction and proceed in the direction of reducing Gibbs free energy. The formation of SiO_x or HfSiO is conducive to reducing Gibbs free energy.

Hereafter, we discuss the oxidation of Ge substrate in ozone. Figure 1.3 shows the GeO_x thickness as a function of ozone oxidation time. The oxidation temperature is in the range from 80°C to 400°C. From Figure 1.3, we can find an increase in the oxidation rate with higher temperature. After 25-min oxidation in ozone, the physical thicknesses of GeO_x are about 2.8, 5.28, 7.8, 9.7 and 12.3 Å at 80°C, 250°C, 300°C, 350°C and 400°C, respectively. Furthermore, an initially linear growth of GeO_x thickness versus time is observed below ~10 s, and then, it becomes parabolic as the oxidation time increases. The two different growth modes suggest that there are two different physical/chemical oxidation mechanisms in the oxidation process. Moreover, the oxidation phenomenon has been well observed

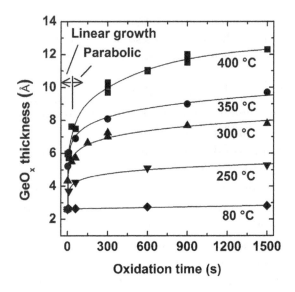

FIGURE 1.3 GeO$_x$ thickness vs. oxidation time at temperatures from 80°C to 400°C.

for Si substrate, which can be effectively interpreted and described by the Deal–Grove or linear parabolic model. As a result, in the region of initially linear growth in Figure 1.3, the oxidation process is determined by chemical reaction, which occurs at the GeO$_x$/Ge interface. However, in the region of parabolic growth, the oxidation is considered to be limited by diffusion process of oxygen atoms through GeO$_x$ film.

In order to accurately well understand the reaction process of Ge oxidation by ozone, the Arrhenius temperature dependence of oxidation process is a good method and measured for each oxidation growth region. Figure 1.4 shows the Arrhenius plot of linear rate constant (B/A) in the initially linear growth region. The B/A is obtained by fitting linear region in Figure 1.3 using the Deal–Grove model. The activation energy is then calculated to be 0.06 eV. This rather small activation energy means that the initially linear growth is nearly barrier-less. And this activation energy is approximately equal to that of Si surface oxidation by ozone. For the parabolic oxidation region, the activation energy is extracted to be 0.54 eV, which is rather reduced compared to the general reported value of thermal oxidation in O$_2$ (~1.7 eV). The small activation energy in ozone oxidation is mainly because of the higher reactivity of oxygen atoms rather than oxygen molecules to diffuse through GeO$_x$, breaking Ge-Ge bond and creating

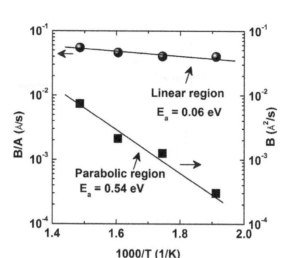

FIGURE 1.4 Arrhenius plots for initially linear region and following parabolic region. The E_a means activation energy.

Ge-O-Ge bond. Therefore, the atomic species during the ozone oxidation induces a reaction pathway, which is much more effective than O_2 species.

1.5 QUANTUM CONFINEMENT EFFECT IN MOS

First, introduce the quantum confinement effect of semiconductor substrate. When the semiconductor substrate is in the strong inversion type, the semiconductor forms a barrier near the interface. This barrier causes the redistribution of the carrier energy level and wave function. It is necessary to re-solve the self-consistent Schrodinger–Poisson equation, and the result is that at the surface, the depletion phenomenon of carriers occurs, and the space range is about 3 Å. This will contribute an additional equivalent oxide thickness of the gate dielectric, which is not conducive to the increase in the capacitance of the gate dielectric (Figure 1.5).

The quantum confinement effect will contribute additional gate-equivalent oxide thickness, reduce the gate capacitance and then reduce the gate control capability and channel current. However, the quantum confinement effect keeps the carriers away from the interface, which can suppress the interface roughness scattering and the remote Coulomb scattering of the gate charge, thereby increasing the mobility.

After the introduction of high dielectric constant gate dielectrics, the concept of equivalent oxide thickness is often used. When extracting the

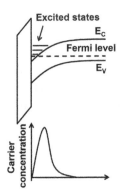

FIGURE 1.5   Quantum confinement in semiconductor inversion layer.

equivalent oxide thickness from the C–V curve of the MOS capacitor, the quantum confinement effect needs to be considered. There are two ways to remove the quantum confinement effect. One is to calculate the equivalent capacitance thickness of the gate structure through the capacitance value of the accumulation zone, and then subtract the quantum confinement effect contribution. For the silicon substrate, usually it is about 3–4 Å. The second method is to solve the self-consistent Schrödinger–Poisson equation of the silicon substrate and solve the self-consistent solution considering the quantum confinement effect, i.e., the relationship between the charge of the silicon substrate and the surface potential. Then, the capacitance–voltage curve of the entire gate structure is simulated. After fitting the experimental curve, the final equivalent oxide layer thickness is obtained. This value has already removed the influence of the quantum confinement effect.

1.6 INTERFACIAL DIPOLE IN MOS GATE STACKS

The interface electric dipole is introduced. The electric dipole moment is a measure of the separation of positive and negative electrical charges in a system of charges, i.e., a measure of the charge system's overall polarity. Figure 1.6 schematically shows positive and negative dipoles at high-κ/ SiO_2 interface. A positive dipole will increase the effective work function (EWF) of metal gate, while a negative dipole will decrease the EWF of the metal electrode as shown in Figure 1.7.

Figures 1.8–1.10 show the investigation on the location of interfacial dipole. The bilayer high-k dielectrics are used to demonstrate if the dipole is located at the high-k/SiO_2 interface. Shown in Figure 1.8 is the V_{FB} shift

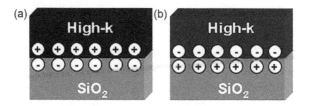

FIGURE 1.6   Schematic of dipole formation at high-k/SiO$_2$ interface. (a) A positive dipole; (b) a negative dipole.

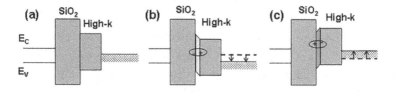

FIGURE 1.7   Schematic of band alignment of gate stacks illustrating the effect of dipole formation at high-k/SiO$_2$ interface on EWF of the metal gate. (a) No dipole; (b) positive dipole; (c) negative dipole.

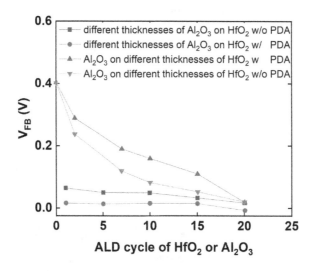

FIGURE 1.8 V_{FB} shift of NiSi/Al$_2$O$_3$/HfO$_2$/SiO$_2$/Si-stacked structure with (w/) or without (w/o) 1000°C PDA (post deposition annealing). PDA is only performed for the bottom-HfO$_2$ layer. V_{FB} shift behavior with PDA is very similar to that without PDA.

of NiSi/Al$_2$O$_3$/HfO$_2$/SiO$_2$/Si stack. It can be seen that the V_{FB} of the gate stack with the top Al$_2$O$_3$ dielectric is approximately the same as that without Al$_2$O$_3$ dielectric. The same results are observed for the NiSi/HfO$_2$/Y$_2$O$_3$/SiO$_2$/Si and NiSi/HfO$_2$/Al$_2$O$_3$/SiO$_2$/Si stacks as shown in Figures 1.9 and 1.10. This means that the top-layer dielectric has no effect on the V_{FB} shift and that the dipole is not located on the metal/high-k and top high-k/bottom high-k interfaces. Inversely, the insertions of bottom-layer high-k dielectric have a significant effect on the V_{FB} shift of the gate stacks. For example, the insertion of HfO$_2$ layer between the Al$_2$O$_3$ and SiO$_2$ layers induces a negative V_{FB} shift of ~0.4 V, shown in Figure 1.8, and the insertion of Y$_2$O$_3$ between HfO$_2$ and SiO$_2$ layers results in a negative V_{FB} shift of ~0.5 V, as shown in Figure 1.9. The introduction of Al$_2$O$_3$ into the HfO$_2$/SiO$_2$ interface makes a positive V_{FB} shift of 0.4 V, as shown in Figure 1.10. It can be concluded that the dipole is located at the high-k/SiO$_2$ interface.

Even though intensive research has been done in the past 5 years about the dipole formation at high-k/SiO$_2$ interface, the exact effect of interfacial dipole on the EWF shift and the physical origin of the dipole formation are still in debate. In this section, the definition of the interfacial dipole will be first discussed, and then the extraction method of dipole moment will be given. Finally, the physical origins of the dipole formation proposed in the literature are investigated.

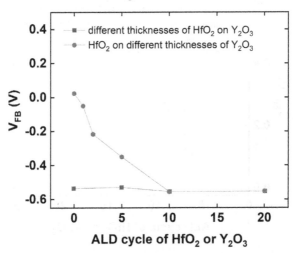

FIGURE 1.9 V_{FB} shift of NiSi/HfO$_2$/Y$_2$O$_3$/SiO$_2$/Si-stacked structure with (w/) or without (w/o) 1000°C PDA. PDA is only performed for the bottom-Y$_2$O$_3$ layer. V_{FB} shift behavior with PDA is very similar to that without PDA.

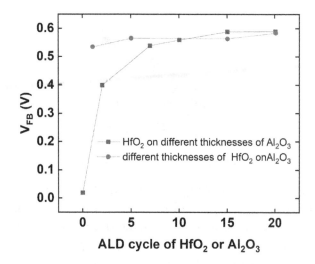

FIGURE 1.10 V_{FB} shift of NiSi/HfO$_2$/Al$_2$O$_3$/SiO$_2$/Si-stacked structure w/ or w/o 1000°C PDA. PDA is only performed for the bottom-Al$_2$O$_3$ layer. The positive V_{FB} shift by about 0.2 V is observed with 1000°C PDA process on Al$_2$O$_3$/SiO$_2$ interface.

1.7 EXTRACTION METHOD OF DIPOLE FORMATION AT HIGH-K/SIO$_2$ INTERFACE

1.7.1 Capacitance–Voltage Method

This method was proposed by our previous work. For MOS capacitors with metal/high-k/terraced-SiO$_2$/Si stack as shown in Figure 1.11, the V_{FB} of this structure is given as follows:

$$V_{FB} = \phi_{ms} - \frac{Q_{SiO_2,Si}\,EOT}{\varepsilon_0\varepsilon_{SiO_2}} - \frac{\rho_{bulk,SiO_2}\,EOT^2}{2\varepsilon_0\varepsilon_{SiO_2}} - \frac{Q_{high\text{-}k,SiO_2}d_{high\text{-}k}}{\varepsilon_0\varepsilon_{high\text{-}k}} - \frac{\rho_{bulk,high\text{-}k}d_{high\text{-}k}^2}{2\varepsilon_0\varepsilon_{high\text{-}k}}$$

$$+ \frac{\rho_{bulk,SiO_2}\varepsilon_{SiO_2}d_{high\text{-}k}^2}{2\varepsilon_0\varepsilon_{high\text{-}k}^2} + \Delta V_{high\text{-}k,SiO_2} + \Delta V_{metal,high\text{-}k} \qquad (1.1)$$

where EOT is the equivalent oxide thickness of the whole metal/high-k/ SiO$_2$/Si stack. ϕ_{ms} is the vacuum work function difference between metal gate and Si substrate. $Q_{SiO_2,Si}$ and $Q_{high\text{-}k,SiO_2}$ are the areal charge densities (per unit area) at SiO$_2$/Si and high-k/SiO$_2$ interfaces, respectively. ρ_{bulk,SiO_2}

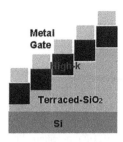

FIGURE 1.11   Schematic of metal/high-k/terraced-SiO$_2$/Si structure.

and $\rho_{bulk,high\text{-}k}$ are the bulk charge densities (per unit volume) in SiO$_2$ and high-k dielectric. $\Delta V_{high\text{-}k,SiO_2}$ and $\Delta V_{metal,high\text{-}k}$ are the V_{FB} shift moments due to the possible dipole at high-k/SiO$_2$ interface and Fermi-level pinning (FLP) at metal gate/high-k interface. ε_0, ε_{SiO_2}, and $\varepsilon_{high\text{-}k}$ express the vacuum permittivity, the relative permittivity of SiO$_2$ and high-k dielectric, respectively. dhigh-k is the physical thickness of high-k dielectric. Then, the intercept (I) of $V_{FB} - EOT$ plot can be obtained to be

$$I = \phi_{ms} - \frac{Q_{high\text{-}k,SiO_2} d_{high\text{-}k}}{\varepsilon_0 \varepsilon_{high\text{-}k}} - \frac{\rho_{bulk,high\text{-}k} d_{high\text{-}k}^2}{2\varepsilon_0 \varepsilon_{high\text{-}k}} + \frac{\rho_{bulk,SiO_2} \varepsilon_{SiO_2} d_{high\text{-}k}^2}{2\varepsilon_0 \varepsilon_{high\text{-}k}^2}$$

$$+ \Delta V_{high\text{-}k,SiO_2} + \Delta V_{metal,high\text{-}k} \qquad (1.2)$$

Then, the EWF of metal gate is expressed as

$$EWF = \phi_m - \frac{Q_{high\text{-}k,SiO_2} d_{high\text{-}k}}{\varepsilon_0 \varepsilon_{high\text{-}k}} - \frac{\rho_{bulk,high\text{-}k} d_{high\text{-}k}^2}{2\varepsilon_0 \varepsilon_{high\text{-}k}}$$

$$+ \frac{\rho_{bulk,SiO_2} \varepsilon_{SiO_2} d_{high\text{-}k}^2}{2\varepsilon_0 \varepsilon_{high\text{-}k}^2} + \Delta V_{high\text{-}k,SiO_2} + \Delta V_{metal,high\text{-}k} \qquad (1.3)$$

It can be concluded that the EWF of the metal gate includes not only the contribution from the vacuum work function of the metal gate, but also the contributions from the following parameters: the areal charge at high-k/SiO$_2$ interface, the bulk charges in the high-k dielectric, the dipole at the high-k/SiO$_2$ interface and the FLP at the metal/high-k interface.

Equation (1.2) shows that I is a quadratic function of $d_{high\text{-}k}$ and that the intercept of I–$d_{high\text{-}k}$ plot is the sum of $\Delta V_{high\text{-}k,SiO_2}$, $\Delta V_{metal,high\text{-}k}$, and ϕ_{ms}. The $\Delta V_{high\text{-}k,SiO_2}$ can be obtained if the $\Delta_{Vmetal, high\text{-}k}$ and ϕ_{ms} are known. The ϕ_{ms}

can be experimentally obtained by considering a control sample with a structure of metal gate/terraced-SiO2/Si, i.e., no high-k dielectric is used. For this case, the V_{FB} is given by

$$V_{FB} = \phi_{ms} - \frac{Q_{SiO_2,Si}\,EOT}{\varepsilon_0\varepsilon_{SiO_2}} - \frac{\rho_{bulk,SiO_2}\,EOT^2}{2\varepsilon_0\varepsilon_{SiO_2}} \qquad (1.4)$$

where ϕ_{ms} is just the intercept of V_{FB}–EOT plot. In addition, $\Delta V_{metal,\,high\text{-}k}$ has been demonstrated to be $0\,V$ for the metal gate without Si component/high-k stacks such as TiN/HfO$_2$, TiN/Al$_2$O$_3$ and TiN/Y$_2$O$_3$ stacks. Thus, $\Delta V_{high\text{-}k,SiO_2}$ can be quantitatively extracted. Furthermore, the $Q_{high\text{-}k,SiO_2}$ and $\rho_{bulk,high\text{-}k}$ can also be calculated through fitting the linear and quadratic parameters of Equation (1.2), respectively.

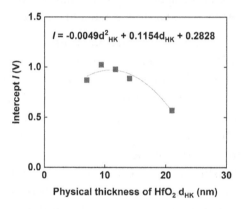

FIGURE 1.12   Extraction of dipole at HfO$_2$/SiO$_2$ interface based on C–V method.

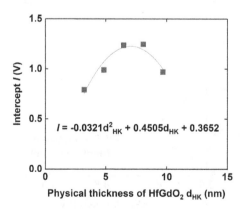

FIGURE 1.13   Extraction of dipole at HfGdO$_x$/SiO$_2$ interface based on C–V method.

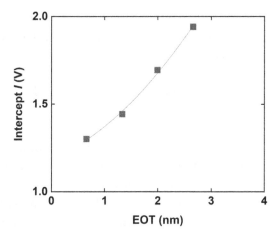

FIGURE 1.14 Extraction of dipole at Al_2O_3/SiO_2 interface based on $C–V$ method. Here, the EOT of Al_2O_3 is used instead of physical thickness of Al_2O_3.

Figures 1.12–1.14 show the extraction of the dipole moment at HfO_2/SiO_2, $HfGdO_x/SiO_2$ and Al_2O_3/SiO_2 interfaces based on this method, respectively. It should be noted that the dipole moment is obtained. This result is obviously different from the reported data by Kamimuta et al. In that case, the $\Delta V_{high\text{-}k,SiO_2}$ is estimated to be about +0.31 V by considering the difference of metal EWF on HfO_2 and SiO_2, as shown in Figure 1.15. And the EWF of the metal gate on HfO_2 is obtained from the linear relationship of V_{FB}–EOT plot for the MOS capacitors with metal/terraced-HfO_2/4-nm-thick SiO_2/Si structure. For the case with terraced-HfO_2 structure, other groups also performed lots of studies.

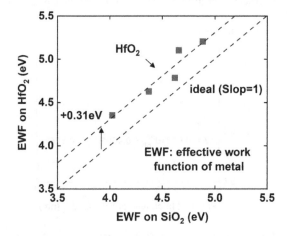

FIGURE 1.15 Extraction of dipole at HfO_2/SiO_2 interface by comparing the EWF shift.

Similarly, however, the effects of dipole and FLP are not considered sufficiently. In this work, a comprehensive relationship of V_{FB}-EOT is expressed as shown in Equation (1.4). By combining Equations (1.3) and (1.4) and considering that $\Delta V_{metal,\ high-k}$ and ρ_{bulk,SiO_2} are negligible, the difference of EWF on HfO_2 and SiO_2 can be given by

$$\Delta EWF = -\frac{\varepsilon_{high-k}\rho_{bulk,high-k}EOT_{SiO_2}^2}{2\varepsilon_0\varepsilon_{SiO_2}^2} + \frac{Q_{high-k,SiO_2}EOT_{SiO_2}}{\varepsilon_0\varepsilon_{SiO_2}} + \Delta V_{high-k,SiO_2} \quad (1.5)$$

It can be seen that the ΔEWF is not only resulted from the $\Delta V_{high-k,SiO_2}$, but also from the $\rho_{bulk,high-k}$, and Q_{high-k,SiO_2}. Thus, the ΔEWF is not simply equal to the $\Delta V_{high-k,SiO_2}$.

1.7.2 Method Based on X-ray Photoemission Spectroscopy

X-ray photoemission spectroscopy (XPS) is widely used to investigate the bonding state of atoms. In addition, XPS can also be employed to demonstrate the band offset of the heterojunction such as high-k/Si contact. The difference of binding energy of core levels can be obtained precisely by the XPS measurement. The changes in difference of binding energy of core levels indicate the band alignment of the whole gate stack. The interfacial dipole formation affects the band alignment of the gate stack, as shown in Figures 1.16–1.18. The effects of positive and negative dipoles on the band alignment

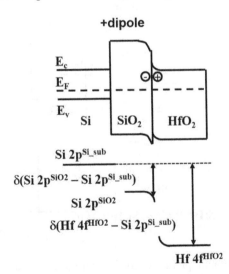

FIGURE 1.16 Schematic picture of the band alignment diagram in $HfO_2/SiO_2/Si$ stacks with a positive interfacial dipole. CB and VB denote the conduction-band and valence-band edges, respectively.

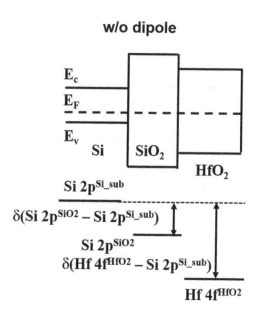

FIGURE 1.17 Schematic picture of the band alignment diagram in SiO_2/Si stacks without an interfacial dipole. CB and VB denote the conduction-band and valence-band edges, respectively.

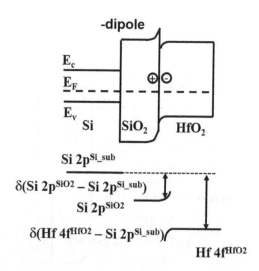

FIGURE 1.18 Schematic picture of the band alignment diagram in $Y_2O_3/SiO_2/Si$ stacks without a negative interfacial dipole. CB and VB denote the conduction-band and valence-band edges, respectively.

are schematically demonstrated. For a positive dipole at high-k/SiO$_2$ interface as shown in Figure 1.16, there is downward band bend for SiO$_2$ at the high-k/SiO$_2$ interface, while upward for HfO$_2$. So the distance of Si 2p core levels between SiO$_2$ and Si substrates increases compared with that with no dipole as shown in Figure 1.17. For a negative dipole at high-k/SiO$_2$ interface as shown in Figure 1.18, upward band bend exists on the SiO$_2$ side, while downward on the high-k dielectric side. So the distance of Si 2p core levels between SiO$_2$ and Si substrates decreases compared with that with SiO$_2$/Si structure as shown in Figure 1.18. Thus, the measurement of Si 2p core levels between SiO$_2$ and Si substrate can be employed to investigate the interfacial dipole formation. Several works have been done to assess the dipole formation at HfO$_2$/SiO$_2$, Al$_2$O$_3$/SiO$_2$, Y$_2$O$_3$/SiO$_2$ and LaAlO$_x$/SiO$_2$ interfaces. The determination of exact magnitudes of dipoles is difficult by XPS measurement. The magnitude of interfacial dipole cannot be directly measured by XPS, and the dipole formation can only be indirectly characterized using the method.

The XPS method is successfully used in characterizing the dipole formation at the high-k/SiO$_2$ interface. In addition, it can give information about band alignment at heterostructures such as the high-k/SiO$_2$ interface. Thus, it is very useful and powerful in investigating the physical origin of the dipole formation at high-k/SiO$_2$ interface. An important disadvantage of this method is the difficulty of correction of charging effect. The charging effect can affect the absolute values of binding energy and make the determination of the core levels of designated atoms unreliable. These result in the limitation in directly investigating the band bend of high-k dielectrics or SiO$_2$.

1.7.3 Method Based on Internal Photoemission

Internal photoemission spectroscopy (IPE) provides the most straightforward way to characterize the relative energies of electron states at interfaces of insulators with metals and semiconductors by measuring the spectral onset of electron/hole photoemission from one solid into another as shown in Figure 1.19. The physical transparency and simplicity of the charge carrier photoemission process are seen to make the IPE a most direct and reliable method of interface barrier characterization.

The dipole magnitude at HfO$_2$/SiO$_2$ interface has been investigated to be +0.3 V. The distance from Fermi level to the conduction band minimum is experimentally to be 2.3 eV for the Al/HfO$_2$/SiO$_2$/Si stack, and

(a)   (b)

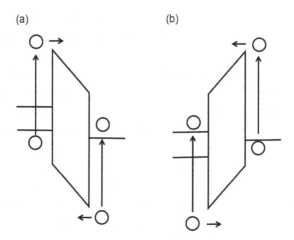

FIGURE 1.19   Photon-excited electron transitions in Si/metal oxide (MeO$_x$)/ metal structures with positive (a) and negative (b) gate voltages.

the barrier height between the valence band maximum of Si substrate and conduction band minimum of HfO$_2$ is 3 eV. Considering the band gap of Si substrate as 1.12 eV, the conduction-band offset between Si substrate and HfO$_2$ is 1.9 eV. Considering that the conduction band offset of Si substrate and SiO$_2$ is 3.15 eV and the electron affinity of SiO$_2$ is 0.9 eV, a potential drop of 0.3 V is found at the HfO$_2$/SiO$_2$ interface, i.e., the magnitude of dipole at HfO$_2$/SiO$_2$ interface is 0.3 V with the positive charges on the HfO$_2$ side.

Similar to XPS measurement, IPE can provide the band alignment of the gate stacks directly. In addition, there is no charging effect in IPE as the XPS measurement. Thus, IPE is a powerful method in assessing the physical origin of dipole formation at high-k/SiO$_2$ interface.

1.8 PHYSICAL ORIGIN OF DIPOLE FORMATION AT HIGH-K/SIO$_2$ INTERFACE

Several models about the physical origin of dipole formation at high-k/ SiO$_2$ interface have been proposed, such as electronegativity model, areal oxygen density model and interface induced gap states model. In this section, the above models are described in detail.

1.8.1 Electronegativity Model

This model uses gate stacks with metal gate/capping layer/HfSiON/SiO$_2$/ Si stack. The capping layers are rare earth (RE)-based high-k dielectrics such as La, Sc, Er and Sr. After 1070°C source/drain activation annealing, the combination of capping layer and the HfSiON occurs, and the Hf(RE) SiON dielectric is formed. Then, the V_{FB} of MOS capacitors with different gate dielectric is measured, and the V_{FB} with the capping layers is more negative than that without capping layers, with the sequence of SrO < Er < Ec+Er < LaO < Sc < none.

Then, a model based on the radii of atoms and the electronegativity is proposed to explain dipole formation at high-k/SiO$_x$ interface. Both the Hf-O bond and RE-O bond form at the high-k/SiO$_x$ interface as shown in Figure 1.20. The Hf-O bond is toward SiO$_x$ interlayer with the positive charges on the high-k dielectric side, while the RE-O bond is toward high-k dielectric with the positive charges on the SiO$_x$ side. The magnitude of interfacial dipole moment is determined by $+Q$ (cation) and $-Q$ (anion), separated by a distance d, given by $D = Q \times d$. The RE-O dipole moment (D_{RE-O}) is larger than the Hf-O dipole moment (D_{Hf-O}) because RE is less electronegative (>Q) and has a larger cationic radius (>d). Thus, the net dipole moment vector points from SiO$_x$ toward high-k dielectric and shifts the metal EWF negative, depending on the dopant type. Both electronegativity and ionic radius increase in the order: Sc < Er < La < Sr. Correspondingly, V_{FB} tuning should be the lowest for Sc and the highest for Sr.

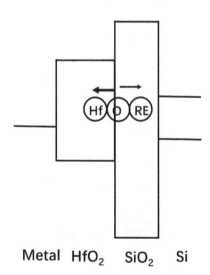

Metal HfO$_2$ SiO$_2$ Si

FIGURE 1.20 Schematic of electronegativity model explaining interface dipole formation and moment.

Even though the electronegativity model successfully explains the above experimental phenomenon, there are some problems about this model.

1. This model cannot explain the negative dipoles at the La_2O_3/SiO_2 and the Y_2O_3/SiO_2 interfaces.

2. The exact depth profile of dopant diffusion needs further determination. Since the dopant is diffused from high-k dielectric to SiO_x interlayer, there should be more dopant on the high-k side, but not on SiO_x side. The electronegativity model, however, presumes that the concentration of dopant is larger on the SiO_x side than on the high-k side.

1.8.2 Areal Oxygen Density Model

In this model, an areal density difference of oxygen atoms at high-k/SiO_2 interface is considered as an intrinsic origin of the dipole formation. The oxygen movement from a higher oxygen density side to a lower oxygen density one will determine the direction of interface dipole as shown in Figure 1.21. There exist positive charges on the oxide side with higher oxygen density, while negative charges are on the oxide side with lower oxygen density. The oxygen densities of HfO_2 and Al_2O_3 are larger than that of SiO_2, while those of Y_2O_3 and La_2O_3 are smaller. So positive dipole will be deduced at HfO_2/SiO_2 and Al_2O_3/SiO_2 interfaces, while negative dipoles at Y_2O_3/SiO_2 and La_2O_3/SiO_2 interfaces. The predicted directions and values of dipoles based on this model are consistent with the experimental values as shown in Figure 1.22.

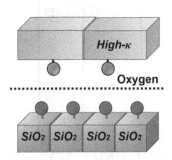

FIGURE 1.21 Schematics of oxygen density model to explain the dipole formation at high-k/SiO_2 interface based on the difference of areal density of oxygen atoms for the case that high-k oxide has smaller oxygen density than SiO_2.

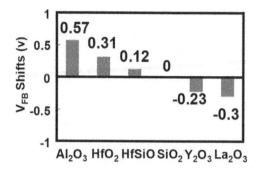

FIGURE 1.22   Experimental results of dipoles at high-k/SiO$_2$ interfaces.

The oxygen density model successfully interprets the direction and values of dipoles at four different high-k/SiO$_2$ interfaces. This model, however, needs a solid theoretical foundation for comprehensive understanding.

1.8.3 Interface Induced Gap States Model

This model discusses the dipole formation based on band alignment of the gate stack. In general, as a high-k dielectric contacts with a SiO$_2$ film closely, the mismatch of two materials causes energy band discontinuity at the interface, which further induces gap states and related charges on both sides. To quantitatively demonstrate the effect of induced gap states and related charges on the electrical properties of MOS device with high-k/metal gate structure, especially on V_{FB} shift, a model of dielectric contact induced gap states (DCIGS) is proposed. The DCIGS model is built on the base of modified metal induced gap state (MIGS) model.

There are two basic models on MIGS, i.e., the fixed separation model and negative charge model. The former assumes that there exists a gap between the metal and the semiconductor, but no spatial distribution of the MIGS charges; the latter assumes that the distribution of the MIGS charges is spatially exponentially extended, but there is no gap.

The DCIGS model considers not only dielectric and dielectric contact, but also the theoretical assumptions of interface gap between two materials and spatially extended exponential distribution of DCIGS charges. The induced gap states are assumed to localize in the forbidden region between conduction band and valence band. The detailed energy band diagram of high-k/SiO$_2$ system based on the DCIGS model is shown in Figure 1.23. ϕ_0 is used to specify the lowest energy level of DCIGS and measured from the valence-band edge of dielectric. The value of ϕ_0 is not given accurately because of the lack of precise calculation data. However, it can be estimated quantitatively by referencing the case of MIGS model. The ϕ_0 in MIGS model is about one-third of Si band gap. Similarly, the ϕ_0 in the DCIGS model is assumed to be about one-third of the dielectric band gap. We consider both high-k and SiO_2 have acceptor surface states whose density is $D_{SS,\,high\text{-}k}$ for high-k and $D_{SS,\,SiO_2}$ for SiO_2. D_{SS} is assumed to be a constant over the energy range from ϕ_0 to Fermi level for both high-k and SiO_2. Q_{SS} is the DCIGS charges. Q_{SC} represents the space charges. $q\Delta$ is the potential drop in the dielectric due to Q_{SS}, and $q\Delta$ is the potential drop across the gap.

In Figure 1.23a, the DCIGS on SiO_2 side are assumed to play a dominant role in the energy band bending of high-k/SiO_2 system compared with that on high-k side, so only $D_{SS,\,SiO_2}$ is considered in Figure 1.23b and $D_{SS,\,high\text{-}k}$ is

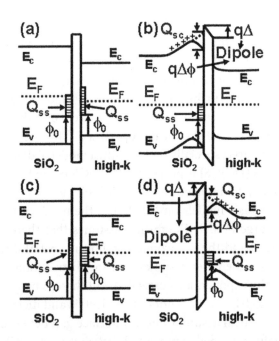

FIGURE 1.23 The detailed energy band diagrams of high-k/SiO_2 systems based on DCIGS model. (a) and (b) correspond to the case that DCIGS on SiO_2 side play a dominant role before contact (a) and at thermal equilibrium after contact (b). (c) and (d) correspond to the case that DCIGS on high-k side play a dominant role before contact (a) and at thermal equilibrium after contact (b). The magnitude of interface dipole is equal to the sum of $q\Delta_1$, $q\Delta_2$ and $q\Delta$.

ignored for analysis simplification. Because the density of DCIGS charges is usually orders of magnitude higher than the space charge density, electrons will flow from high-k to SiO_2. So negative charges are built up on SiO_2 side, and equal positive charges on high-k side. Considering the gap between high-k and SiO_2 and spatially extended exponential distribution of Q_{SS}, the detailed energy band diagram at thermal equilibrium is shown in Figure 1.23b. There occurs a peak in the energy band diagram due to the exponential distribution of Q_{SS}. A positive potential drop crosses the interface from high-k to SiO_2. Consequently, a positive dipole at high-k / SiO_2 interface is induced. On the other hand, if $D_{SS,}{}^{high\text{-}k}$ plays a dominant role and D_{SS,SiO_2} is ignored, a negative dipole formation can be demonstrated as shown in Figure 1.23c and d.

Furthermore, magnitude of dipole is studied from the viewpoint of band alignment of high-k/SiO_2 systems by resolving the Poisson's equation. In Figure 1.23b, the bulk density of DCIGS charges on SiO_2 side ρ_{SS} is assumed to decay exponentially into SiO_2, and ρ_{SS} is given by

$$\rho_{SS} = \frac{Q_{SS}}{\lambda} \exp\left(-\frac{z}{\lambda}\right) \tag{1.6}$$

where Q_{SS}, the areal density of the net DCIGS charges on SiO_2 side, is negative because acceptor DCIGS have been assumed above. λ is the penetration depth of DCIGS. The z axis is defined normal to high-k/SiO_2 interface. The zero point lies to SiO_2 surface, and positive direction points to internal SiO_2. Q_{SS} is given by

$$Q_{SS} = -eD_{SS,SiO_2}(E_F - \phi_0) \tag{1.7}$$

So the Poisson's equation on SiO_2 side is expressed as

$$\frac{d^2\phi}{d^2z} = -\frac{Q_{SS}}{\varepsilon_0\varepsilon_{SiO_2}\lambda} \exp\left(-\frac{z}{\lambda}\right) \tag{1.8}$$

And $q\Delta\phi$ is calculated to be

$$q\Delta\phi = -\frac{\lambda Q_{SS}}{\varepsilon_0\varepsilon_{SiO_2}} \tag{1.9}$$

$q\Delta$ is equal to

$$q\Delta = -\frac{\delta_{gap} Q_{SS}}{\varepsilon_0 \varepsilon_{int}} \tag{1.10}$$

δ_{gap} and ε_{int} are the thickness and permittivity of the gap, respectively. So the magnitude of the positive interface dipole is expressed as

$$\Delta V = \frac{e\lambda D_{SS,SiO_2}(E_F - \phi_0)}{\varepsilon_0 \varepsilon_{SiO_2}} + \frac{e\delta_{gap} D_{SS,SiO_2}(E_F - \phi_0)}{\varepsilon_0 \varepsilon_{int}} \tag{1.11}$$

Similarly, the magnitude of negative interface dipole in Figure 1.23d is given by

$$\Delta V = -\frac{e\lambda D_{SS,SiO_2}(E_F - \phi_0)}{\varepsilon_0 \varepsilon_{high-k}} - \frac{e\delta_{gap} D_{SS,high-k}(E_F - \phi_0)}{\varepsilon_0 \varepsilon_{int}} \tag{1.12}$$

Furthermore, the CNL is employed for the first time to extract the direction and magnitude of the dipole at high-k/SiO$_2$ stacks. The DCIGS model discusses the dipole formation from the energy band alignment of the high-k and SiO$_2$ using the DCIGS and bulk states. It is because that the DCIGS and bulk states determine the energy band alignment of the high-k and SiO$_2$ contact. The Fermi levels are consistent for the high-k and SiO$_2$ contact at the thermal equilibrium. And the charge transfer between the high-k and SiO$_2$ is deduced by the requirement of the energy band alignment. Considering the definition of the interface dipole, it is concluded that the physical origin of the dipole formation should be demonstrated from the viewpoint of the band alignment, which is the DCIGS model for the high-k and SiO$_2$ contact.

The DCIGS theory for high-k and SiO$_2$ contact can be considered to be an extension of the gap states concept, which is initially used in discussing the energy band structures of metal and semiconductor contact or semiconductor and insulator contact. The MIGS is usually used to interpret the metal and semiconductor contact. A similar approach for contact between two semiconductors or between semiconductor and insulator is proposed by using CNL in analogy with Fermi level. The high-k dielectric and SiO$_2$ can be regarded as wide bandgap semiconductors. Naturally, we can employ the DCIGS and CNL concepts to describe and determine the energy band structure lineup for high-k dielectric and SiO$_2$ contact.

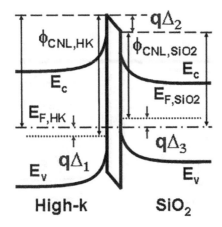

FIGURE 1.24 The schematic diagram of energy band structure for a negative dipole formation in a high-k/SiO$_2$ system.

For the high-k and SiO$_2$ contact, the DCIGS that derives from the virtual gap states of the complex band structure of the dielectrics is presumably considered to exist on both sides of the two dielectrics because of the spatial interruption of their respective atom distribution structures. The DCIGS is an intrinsic characteristic of the dielectric and consists of the valence- and conduction-band states. The characteristic of the DCIGS changes across the band gap from predominately donor- to acceptor-like closer to the valence band top and the conduction band bottom, respectively. The energy at which their characteristic changes is called their branch point, or most generally, CNL. The CNL plays a role of the Fermi level.

The schematic diagram of the energy band structure lineup for the case of a negative dipole formation in a high-k/SiO$_2$ system is shown in Figure 1.24. The continuum of the DCIGS determines the energy band structure at high-k/SiO$_2$ interface. As the CNL of the high-k is lower than that of the SiO$_2$ before dielectrics contact, there occurs electron transfer from the SiO$_2$ to high-k in order to decrease the energy difference of the CNLs between high-k and SiO$_2$, or namely, balance the Fermi levels of these two dielectrics. Consequently, there exist net positive charges on the SiO$_2$ side, while equal negative charges on the high-k side, which results in a voltage drift at high-k/SiO$_2$ interface from the SiO$_2$ to high-k, i.e., a negative interface dipole is formed. On the other hand, if the CNL of the high-k is higher than that of the SiO$_2$ before

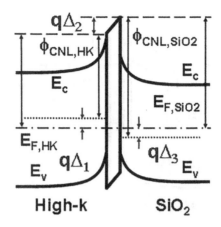

FIGURE 1.25 The schematic diagram of energy band structure for a positive dipole formation in a high-k/SiO$_2$ system.

dielectric contact, there occurs electron transfer from the high-k to the SiO$_2$, which implies a positive dipole formation at high-k/SiO$_2$ interface as depicted in Figure 1.25.

Furthermore, the strength of the electric dipole at high-k/SiO$_2$ interface is quantitatively studied from the viewpoint of energy band structure lineup. As shown in Figure 1.24, the CNL of the high-k is lower than the Fermi level due to the excessive electrons on the high-k side, while the CNL of the SiO$_2$ is higher than the Fermi level due to the insufficient electrons on the SiO$_2$ side, implying that there appear voltage drifts Δ_1 and Δ_3 on the high-k and SiO$_2$ sides, respectively. And the Δ_1 and Δ_3 can be expressed by

$$\Delta_1 = \frac{E_{F,HK} - \phi_{CNL,HK}}{q} \tag{1.13}$$

$$\Delta_3 = \frac{\phi_{CNL,SiO_2} - E_{F,SiO_2}}{q} \tag{1.14}$$

$E_{F,HK}$, E_{F,SiO_2}, $\phi_{CNL,HK}$ and ϕ_{CNL,SiO_2} express the Fermi levels and CNLs of high-k and SiO$_2$ with respect to their vacuum levels, respectively. In addition, there also appears a voltage drift Δ_2 on the spatial gap between the high-k and SiO$_2$ dielectrics. The Δ_2 can be obtained by

$$\Delta_2 = \frac{E_{F,SiO_2} - E_{F,HK}}{q} \tag{1.15}$$

It is noted that the Fermi levels of both the high-k and SiO_2 are consistent and that the different Fermi levels of high-k and SiO_2 after dielectric contact are due to the different reference energy levels defined here. As a result, the strength of the electric dipole at high-k/SiO_2 interface can be expressed as the sum of Δ_1, Δ_2 and Δ_3 as follows:

$$\Delta = \Delta_1 + \Delta_2 + \Delta_3 \qquad (1.16)$$

From Equations (1.13)–(1.16), the dipole moment Δ can be given by

$$\Delta = \frac{\phi_{CNL,SiO_2} - \phi_{CNL,hk}}{q} \qquad (1.17)$$

It can be seen that the dipole strength at high-k/SiO_2 interface is equal to the difference of CNLs between high-k and SiO_2, which can be used to quantitatively calculate the dipole moments. Similarly, the above discussions can also be extended to the high-k/SiO_2 systems with a positive dipole as demonstrated in Figure 1.25.

The DCIGS model and CNL concept explain the dipole formation at high-k/SiO_2 interface from band alignment of the gate stack. Thus, it provides a more comprehensive physical insight into the dipole formation. In addition, this model can quantitatively predict the values of dipoles at high-k/SiO_2 interface.

1.9 "ROLL-OFF" AND "ROLL-UP" PHENOMENON

One challenge for obtaining EWF near the valence-band edge of Si substrate is the V_{FB} Roll-Off phenomenon. When the metal/high-k/terraced-SiO_2/Si structure is used to extract EWF of the metal gate, the relationship between the V_{FB} and EOT should be linear. However, when the thickness of interlayer SiO_2 decreases below ~4 nm, the slope of V_{FB}–EOT curve shows a strong roll-off phenomenon. The roll-off phenomenon is found using high-k dielectrics such as HfO_2 and Al_2O_3. Roll-off phenomenon makes the tuning of EWF for PMOS (p type metal-oxide-semiconductor) negative and difficult.

On the other hand, V_{FB}-EOT roll-up phenomenon is reported, i.e., the V_{FB}–EOT curves start to roll up when the interfacial layer SiO_2 is decreased below ~4 nm. When the interlayer SiO_2 continues to decrease, the roll-off phenomenon appears.

The tuning of band edge EWF is one of the main challenges for the MOS device with high-k/metal gate structure. It is found that there occurs flat-band voltage (V_{FB}) roll-off when the interfacial layer SiO_2 between the high-k dielectric and Si substrate becomes sufficiently thin (~3 nm), which is particularly negative for the threshold voltage tuning of the PMOS. For the origin of the V_{FB} roll-off, there are several interpretations. Bersuker et al. suggested that it was associated with the generation of positively charged oxygen vacancies in the interfacial layer SiO_2 next to the Si substrate. Choi et al. considered that the dipole at the high-k/SiO_2 interface is the origin of the V_{FB} roll-off. Zheng et al. reported that the dipole at SiO_2/ Si interface due to the O atom diffusion caused the V_{FB} roll-off.

Here a possible physical origin of the V_{FB} roll-off phenomenon is proposed based on the consideration of the contact of the high-k dielectric and Si substrate, i.e., from the viewpoint of the electron transfer due to the energy band alignment of the high-k/Si interface. Quantitatively calculated simulation results based on this model are consistent with the experimental data.

When the interfacial layer SiO_2 becomes sufficiently thin, the quantum mechanical tunneling between the high-k dielectric and Si substrate is possible. Charge loss via this tunneling mechanism is generally observed for metal-insulator-SiO_2-silicon nonvolatile memory structures with thin SiO_2 thickness. As a result, the interfacial layer SiO_2 will be transparent to electrons and the contact and interaction of high-k dielectric and Si substrate, or namely, the thermal equilibrium of the high-k with the Si interface, should be considered for the V_{FB} expression when the interface layer SiO_2 is thin enough. And the interfacial layer SiO_2 can be regarded as the gap material between the high-k dielectric and Si substrate. Consequently, it is a high-k/Si system with the gap dielectric constant being 3.9, which is different with the case that the gap material is considered as vacuum and the dielectric constant is 1. The fundamental theory, however, is likewise for these two systems because they are both of oxide/semiconductor contact. Thus, the conventional theory for the energy band alignment of the oxide and semiconductor contact is applied to the high-k/Si interface with a sufficiently thin (≤~3nm) gap layer of SiO_2.

Then, the effect of the high-k/Si interface on the V_{FB} of the MOS with metal/high-k/thin SiO_2/Si structure is discussed based on the energy band alignment of the high-k and Si. For simplification, herein various charges in the metal/high-k/SiO_2/Si stack are ignored and the vacuum work function of metal gate is assumed to be the same as that of the Si substrate. When the interfacial layer SiO_2 is thick, i.e., there is no V_{FB} roll-off, the thermal equilibrium between the high-k and Si need not be considered and the V_{FB} can be easily obtained to be zero. However, when the SiO_2 is sufficiently thin, the thermal equilibrium between the high-k and Si should be considered. Before the contact of the whole system containing metal, high-k, interfacial layer SiO_2 and Si substrate, the system is not in thermal equilibrium and the energy bands are flat as shown in Figure 1.26a. If the Fermi level of high-k dielectric is assumed to be higher than that of Si, after contact of the whole system, electrons will flow from high-k to Si in order to balance their Fermi levels, i.e., the high-k/Si interface is at the thermal equilibrium. Thus, there occurs energy band upward bend on the high-k side, while downward bend on the Si side. Then, net positive charges are built up on the high-k side, while equal and net negative charges on the Si side. Relative to the Fermi level in the high-k dielectric, the Fermi level in the Si substrate is raised by an amount equal to the difference between the two work functions as shown in Figure 1.26b. In addition, the density of the interface gap states on Si substrate is low enough to be ignored for the region of the V_{FB} roll-off. For the case of the

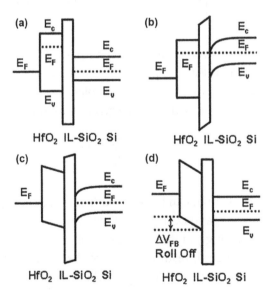

FIGURE 1.26 The schematic diagram of energy band structure for a metal/high-k/SiO_2/Si system showing origin of V_{FB} roll-off. (a) before contact, (b) after contact of high-k, SiO_2 and Si, (c) after contact with metal, and (d) the roll-off.

high-k dielectric, the density of the interface gap states is assumed to be sufficiently large to accommodate any additional surface charges resulting from the alignment of the Fermi levels. So the space charges in the high-k dielectric will remain nearly unaffected and the energy band unchanged. Then, the whole voltage drop falls through the Si substrate and the interfacial gap layer SiO_2 between the high-k dielectric and the Si substrate, as shown in Figure 1.26b. And we can obtain

$$\varphi_S + \frac{\text{EOT}_{SiO_2}}{\varepsilon_0 \varepsilon_r}|Q| = \Delta E_F \tag{1.18}$$

Here, φ_S is the surface potential of Si and it is defined as positive when the energy band is bent downward. ε_0 and ε_r are the permittivity of the vacuum and SiO_2, respectively. EOT_{SiO_2} is the equivalent oxide thickness of SiO_2. ΔE_F is the Fermi-level difference of the high-k dielectric and Si substrate, i.e., the difference of their work functions. Q is the space charge in the Si substrate, and Q can be expressed as follows:

$$|Q| = \frac{\sqrt{2}\varepsilon_0 \varepsilon_S \kappa T}{qL_D}\left[\left(e^{-\beta\varphi_S} + \beta\varphi_S - 1\right) + \frac{n_0}{p_0}\left(e^{-\beta\varphi_S} - \beta\varphi_S - 1\right)\right]^{\frac{1}{2}} \tag{1.19}$$

Here k is the Boltzmann constant. ε_S is the permittivity of Si substrate. T is the temperature in Kelvin. q is the electron charge. n_0 and p_0 are the equilibrium densities of the electrons and holes, respectively. β is equal to q/kT. L_D is the Debye length and can be given as follows:

$$L_D = \sqrt{\frac{\varepsilon_0 \varepsilon_S}{qP_0\beta}} \tag{1.20}$$

Then, the surface potential can be resolved from Equations (1.18) to (1.20) after giving ΔE_F, N_0, P_0 and EOT_{SiO_2}. Furthermore, the space charge can be obtained by substituting φ_S into Equation (1.18).

After the thermal equilibrium contact of the high-k dielectric and Si substrate, the Fermi level of Si is higher than metal gate and electrons will flow from Si substrate to the metal gate in order to balance the Fermi levels, as shown in Figure 1.26c. The energy band of the Si substrate for this case, however, is not flat but bent downward. This is due to the fact that the voltage drop in order to balance the Fermi levels of the metal gate and Si substrate falls on the high-k, interfacial layer SiO_2 and Si substrate. This is different from the thermal equilibrium between high-k dielectric and Si substrate, where the voltage drop in order to balance the Fermi levels of the high-k and Si substrate falls on only the interfacial layer SiO_2 and Si substrate but no high-k dielectric. The flat-band voltage needed for flat-band condition of the Si substrate shown in Figure 1.26d can be obtained to be

$$V_{FB} = -\frac{EOT_{high\text{-}k} + EOT_{SiO_2}}{\varepsilon_0 \varepsilon_r}|Q| - \varphi_S + \Delta E_F = -\frac{EOT_{high\text{-}k}}{\varepsilon_0 \varepsilon_r}|Q| \quad (1.21)$$

Here, $EOT_{high\text{-}k}$ is the equivalent oxide thickness of high-k dielectric. For the conventional case where the thickness of the interfacial layer SiO_2 is thick enough, the flat-band voltage is zero because the same work functions of the high-k and Si have been supposed. Finally, we can define the degree of the V_{FB} roll-off as

$$\Delta V_{FB}\text{roll-off} = \text{actual } V_{FB} - \text{expected } V_{FB} \quad (1.22)$$

ΔV_{FB} roll-off is defined by taking the value of the V_{FB} from the actual region of roll-off and then subtracting the expected V_{FB} obtained by extrapolating the V_{FB}-EOT trend line from the thicker region of the SiO_2 where there is no V_{FB} roll-off occurring. As a result, the ΔV_{FB} roll-off based on this model can be finally obtained to be

$$\Delta V_{FB} \text{ roll-off} = -\frac{EOT_{high\text{-}k}}{\varepsilon_0 \varepsilon_r}|Q| \quad (1.23)$$

It should be noted that although Equation (1.23) is derived based on the assumption that the vacuum work function of the metal gate is consistent with Si substrate, the same equation can also be given when their Fermi levels are different.

Then, the ΔV_{FB} roll-off is calculated based on the proposed model. The experimental results and the simulation data are shown in Figure 1.27. It can be seen that the simulation results are in agreement with the experimental data with the error of the ~0.05 V. The errors may come from the reasons as follows:

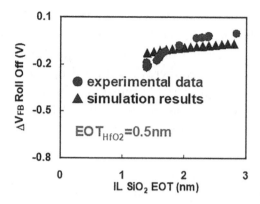

FIGURE 1.27 The V_{FB}–EOT plots of the experimental and simulation results. The doping concentration of experimentally used n-Si substrate is $5\times10^{18}\,cm^{-3}$.

1. The effect of the fixed charges in the metal/high-k/SiO_2/Si stack.

2. The O vacancy induced in the interfacial layer SiO_2.

However, it is significant that this model successfully predicts the trend of the V_{FB} roll-off when the interfacial layer SiO_2 becomes sufficiently thin.

In addition, the effect of thickness of the high-k dielectric, the doping concentration of Si substrate and the doping type are also simulated based on this model. It can be seen that the V_{FB} roll-off increases with increasing thickness of the high-k dielectric as shown in Figure 1.28. The

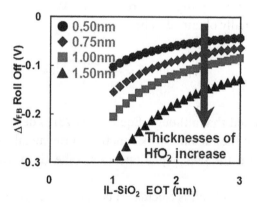

FIGURE 1.28 The simulated V_{FB}–EOT plot for different thicknesses of high-k dielectric.

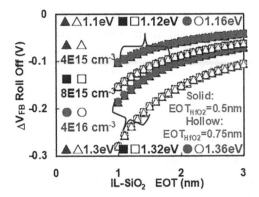

FIGURE 1.29 The simulated V_{FB}–EOT plot for different doping concentration of Si substrate. The corresponding differences of the Fermi levels used in the simulation are simultaneously given at the top and bottom of the figure.

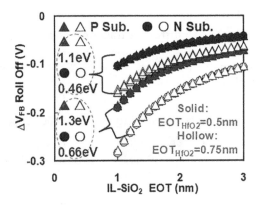

FIGURE 1.30 The simulated V_{FB}–EOT plot for different doping type of Si substrate. The corresponding differences of the Fermi levels used in the simulation are simultaneously given at the top and bottom of the figure.

V_{FB} roll-off is independent of the doping concentration of Si substrate as shown in Figure 1.29 and is also independent of the doping types as shown in Figure 1.30.

In conclusion, a physical model based on the energy band alignment of the high-k/Si interface is proposed to investigate the origin of the V_{FB} roll-off phenomenon. The V_{FB} roll-off is assigned to associate with the thermal equilibrium contact of high-k dielectric and Si substrate. When the interfacial layer SiO_2 becomes sufficiently thin, electrons will transfer between high-k dielectric and Si substrate in order to reach thermal equilibrium, or namely, balance their Fermi levels. Additional negative voltage must apply to approach the flat-band condition of the Si substrate for the whole stack, which induces the V_{FB} roll-off. The calculated results are consistent with the experimental data. It will be very helpful in understanding the physical mechanism of the V_{FB} roll-off and further engineering the V_{FB} of MOS device with high-k/metal gate structure.

1.10 PHYSICAL ORIGIN OF FIXED CHARGES AT GE/GEO$_X$ INTERFACE

First, we discuss the fixed charges at Ge/GeO$_x$ interface. Three sets of samples were used in the experiment to study the origin of charge at the Ge/GeO$_x$ interface. Using the GeO$_x$/Al$_2$O$_3$-laminated film structure, explore the characteristics of GeO$_x$ through XPS. The gate capacitance structure of Ge/GeO$_x$/Al$_2$O$_3$/Al is used to analyze the charge characteristics of the

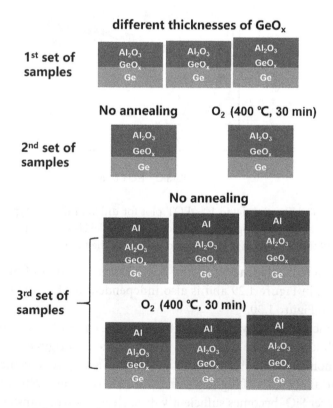

FIGURE 1.31   Schematic of experimental samples.

Ge/GeO$_x$ interface from the perspective of electrical testing. The three sets of experimental samples are shown in Figure 1.31. The first set of samples were oxidized with different thicknesses of GeO$_x$ on Ge substrate by ALD (atomic layer deposition) process. The growth temperature of GeO$_x$ was 300°C, and the oxidation time was 1, 60, 150, 300 and 1500 s, respectively. Subsequently, 2 nm Al$_2$O$_3$ was grown using TMA (Trimethylaluminums) and H$_2$O. Finally XPS were measured. Al$_2$O$_3$ on GeO$_x$ acts as a protective layer to prevent GeO$_x$ from being oxidized in the air. For the second set of samples, the Ge/GeO$_x$/Al$_2$O$_3$ samples with an oxidation time of 1500 s were annealed in an O$_2$ atmosphere at a temperature of 400°C for 30 min and then subjected to XPS testing. The third group of samples is a Ge/GeO$_x$/Al$_2$O$_3$ capacitor structure. Ozone oxidation was performed at 300°C, 350°C and 400°C for 1500 s to grow GeO$_x$ with different thicknesses. Here, the thickness of GeO$_x$ is controlled by different temperatures to ensure a good Ge/GeO$_x$ interface and reduce the interface state density. 10 nm Al$_2$O$_3$ was deposited on GeO$_x$. A set of Ge/GeO$_x$/Al$_2$O$_3$ structures with three different GeO$_x$ thicknesses were used as comparative samples. Al electrodes were deposited directly without any treatment. Another group of Ge/GeO$_x$/Al$_2$O$_3$ structures with three different GeO$_x$ thicknesses were annealed in O$_2$ atmosphere at 400°C for 30 min, and then Al electrode was deposited.

The thickness of GeO$_x$ under different oxidation time is quantitatively calculated by XPS measurement of the ratio of GeO$_x$ and Ge 3d signal intensity in Ge substrate. A photoelectron with an initial intensity I_0 is emitted to the surface at a certain position below the surface of the film, and the expression of the intensity I of the photoelectron reaching the surface is:

$$I = I_0 \cdot e^{-\frac{t}{\lambda}} \tag{1.24}$$

The t is the distance between the atom emitting photoelectrons and the surface, and λ is the mean free path of photoelectrons. For Ge/GeO$_x$ structure, the signal area intensity ratio of GeO$_x$ to Ge 3d in Ge (I_{GeO_x}/I_{Ge}) is:

$$\frac{I_{GeO_x}}{I_{Ge}} = \frac{X_{GeO_x}}{X_{Ge}} \cdot \frac{\lambda_{GeO_x}}{\lambda_{Ge}} \left(e^{\frac{d}{\lambda_{GeO_x} \cdot \sin\theta}} - 1 \right) \tag{1.25}$$

X_{GeOx} and X_{Ge} are the atomic densities of Ge in GeO_x and Ge substrates, respectively; λ_{GeOx} and λ_{Ge} are the inelastic mean free paths of photoelectrons of Ge 3d passing through GeO_x and Ge substrates; d is the thickness of GeO_x film; and θ is photoelectron combined with the high-resolution transmission electron microscopy (HRTEM) measurement of 0.7 nm GeO_x sample, the value of $(X_{GeOx}/X_{Ge}) \cdot (\lambda_{GeOx}/\lambda_{Ge})$ is calculated to be 0.94. The calculation expression of GeO_x thickness is as follows:

$$d = \lambda_{GeO_x} \cdot \sin\theta \cdot \ln\left(\dfrac{\dfrac{I_{GeO_x}}{I_{Ge}}}{\dfrac{X_{GeO_x}}{X_{Ge}} \cdot \dfrac{\lambda_{GeO_x}}{\lambda_{Ge}}} + 1 \right) \qquad (1.26)$$

The GeO_x thickness changes for different oxidation times calculated by the above method are shown in Figure 1.32. From the growth curve, we can see two distinct growth processes with different trends, from the initial linear growth to the subsequent parabolic growth.

Figure 1.33 shows the Ge 3d XPS spectra with the oxidation time of 60 and 1500 s. As the oxidation time increases, the thickness of GeO_x increases, and the Ge 3d signal from GeO_x increases. The samples with the oxidation time of 60 and 1500 s all have Ge^{1+} signals, indicating that the Ge surface is not sufficiently oxidized in the initial oxidation stage. The GeO_x generated by the 60 s oxidation time has almost no Ge^{4+} signal, while for the 1500 s oxidation time, a clear Ge^{4+} signal is observed. As the oxidation time progresses, the surface of GeO_x is more fully oxidized and high-valent germanium oxide GeO_2 is generated.

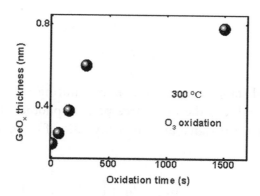

FIGURE 1.32   The GeO_x thickness vs time at 300°C.

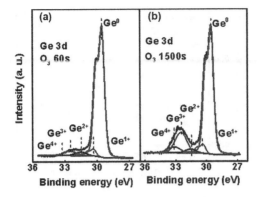

FIGURE 1.33 The XPS spectra of Ge 3d for 60 and 1500 s oxidation.

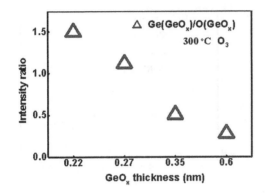

FIGURE 1.34 The intensity ratio of Ge to O for different GeO_x thicknesses.

Figure 1.34 reflects the change of the ratio of Ge to O content in GeO_x. As the thickness of GeO_x increases, the ratio of $Ge(GeO_x)$ to $O(GeO_x)$ gradually decreases. This indicates that when GeO_x is relatively thin, the content of O in GeO_x is relatively low, and when the thickness of GeO_x increases, the content of O in GeO_x also increases, and the bottom region of GeO_x has a lower content of oxygen than the top region. Figure 1.35 shows the changes in the ratio of different Ge oxidation states (Ge^{1+}, Ge^{2+}, Ge^{3+}, Ge^{4+}) to Ge^0 in different thicknesses of GeO_x, which reflects the absolute changes in each oxidation state of Ge. Ge^{1+} is almost stable and does not change with the increase of GeO_x thickness. The content of Ge^{2+} increases slowly with the increase of the thickness of GeO_x. Ge^{3+} has an obvious increasing trend, and Ge^{4+} has a significant increase when the thickness of GeO_x increases to 0.35 nm. Figure 1.36 reflects the relative change of different Ge oxidation states (Ge^{1+},

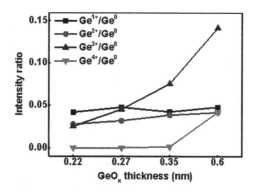

FIGURE 1.35 The changes in the ratio of different Ge oxidation states (Ge^{1+}, Ge^{2+}, Ge^{3+}, Ge^{4+}) to Ge^0 in different thicknesses of GeO_x.

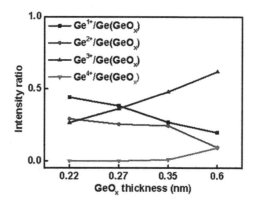

FIGURE 1.36 The relative change of different Ge oxidation states (Ge^{1+}, Ge^{2+}, Ge^{3+}, Ge^{4+}) in GeO_x, i.e., the oxidation state of each Ge and the total Ge content in GeO_x ($Ge^{1+}+Ge^{2+}+Ge^{3+}+Ge^{4+}$) ratio.

Ge^{2+}, Ge^{3+}, Ge^{4+}) in GeO_x, i.e., the oxidation state of each Ge and the total Ge content in GeO_x ($Ge^{1+}+Ge^{2+}+Ge^{3+}+Ge^{4+}$) ratio. As the thickness of GeO_x increases, the relative content of Ge^{1+} and Ge+ in GeO_x decreases, while the relative content of Ge^{3+} and Ge^{4+} in GeO_x increases. Therefore, when the oxidation time increases, the thickness of GeO_x increases, the upper region is oxidized more fully, and the content of high-valent oxidation states gradually increases. The oxidation state of Ge in GeO_x gradually transitions from a low oxidation state to a high oxidation state, and insufficient oxidation causes oxygen vacancies at the Ge/GeO_x interface.

Another way to verify oxygen vacancies is to observe the effect of O_2 annealing on the Ge/GeO_x interface. Figure 1.37a and b show the ratio of

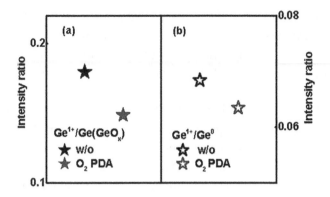

FIGURE 1.37 (a) The ratio of Ge^{1+} to GeO$_x$ in Ge content before and after annealing, (b) the change of the ratio with Ge^0 in Ge substrate.

Ge^{1+} to GeO$_x$ in Ge content before and after annealing, and the change of the ratio with Ge^0 in Ge substrate. The results show that after O$_2$ annealing, the overall change of Ge^{1+} is compared with that in GeO$_x$. The relative changes are reduced. The formation of Ge^{1+} is mainly caused by insufficient oxidation. Therefore, O$_2$ annealing will change the content of Ge^{1+}.

When O$_2$ is not sufficiently oxidized, it will cause oxygen vacancies and form a transition layer of suboxide on the Ge surface. According to the density functional theory (DFT), oxygen vacancies in the +2 charge state exhibit the lowest formation energy. Robertson used DFT to study

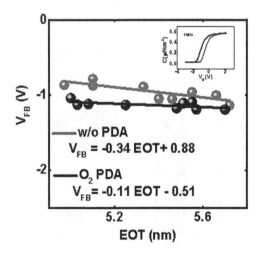

FIGURE 1.38 The V_{FB}–EOT curves before and after the O$_2$ annealing.

the oxidation mechanism of Ge surface and mentioned that the existence of oxygen vacancies on the Ge surface is mainly due to its relatively low formation energy. Through the O_2 annealing and non-annealing capacitor test, the amount of charge at the Ge/GeO$_x$ interface is extracted, and the result is shown in Figure 1.38.

For MOS capacitors with a Ge/GeO$_x$/Al$_2$O$_3$ structure, the relationship between the flat-band voltage and EOT has deviated from the O_2 annealing sample compared with the non-annealed sample. When the thickness of GeO$_x$ is different, the interface charge quantity of Ge/GeO$_x$ can be extracted through the relationship between flat-band voltage and EOT. The relationship between V_{FB} and EOT is restated as follows:

$$V_{FB} = -\frac{Q_1}{\varepsilon_0 \varepsilon_r} \text{EOT} - \frac{\varepsilon_1 \rho_1}{2\varepsilon_0 \varepsilon_r^2} \text{EOT}^2$$

$$- \frac{Q_2}{\varepsilon_0 \varepsilon_r} \text{EOT}_2 + \frac{(\varepsilon_1 \rho_1 - \varepsilon_2 \rho_2)}{2\varepsilon_0 \varepsilon_r^2} \text{EOT}_2^2 + \Delta + \phi_{ms} \qquad (1.27)$$

Q_1 represents the charge density at the Ge/GeO$_x$ interface. The fitting curve of V_{FB} and EOT in Figure 1.38 is linear, and the slope of the curve represents the value of the Ge/GeO$_x$ interface charge Q_1. Compared with the non-annealed sample, the slope of the fitted curve of V_{FB} and EOT decreases after oxygen annealing. The inset is the C–V curve of MOS capacitor before and after O_2 annealing. The shapes of the two curves are almost the same, so O_2 annealing does not affect the interface state of the Ge/GeO$_x$ interface, but the C–V curve after O_2 annealing is obviously shifted to the left, which also reflects the change of the fixed charge at the

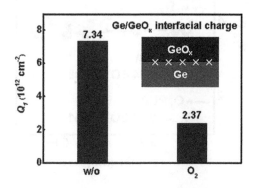

FIGURE 1.39 The GeO$_x$/Ge interfacial charges w/ and w/o O_2 annealing.

interface. The quantitative extraction of the charge quantity at the Ge/GeO_x interface is shown in Figure 1.39. The charge quantity before O_2 annealing is $+7.34 \times 10^{12} cm^{-2}$; after annealing, it decreases to $+2.37 \times 10^{12} cm^{-2}$. The oxygen vacancies caused by insufficient oxidation at the Ge/GeO_x interface are well passivated in the O_2 atmosphere annealing.

1.11 SUMMARY

This chapter introduces the physical origin and passivation of the interface state of MOS devices. Different semiconductor substrates have different sources of interface state and should be treated differently, and the passivation method should be changed accordingly. In general, the interface state is generated because the energy level in the conduction band or the valence band is disturbed and enters the forbidden band gap, showing the characteristics of the interface state. The passivation method of the interface state is very dependent on the in-depth understanding of interface thermodynamics/thermodynamics. The interfacial dipole at high-k/SiO_2 interface is discussed, including its formation, measurement and physical origin.

BIBLIOGRAPHY

1. Ogawa, S. and Y. Takakuwa, Interfacial oxidation kinetics at $SiO_2/Si(001)$ mediated by the generation of point defects: Effect of raising O_2 pressure. *AIP Advances*, 2018. **8**(7): p. 075119.
2. Ogawa, S. and Y. Takakuwa, Rate-limiting reactions of growth and decomposition kinetics of very thin oxides on Si(001) surfaces studied by reflection high-energy electron diffraction combined with auger electron spectroscopy. *Japanese Journal of Applied Physics*, 2006. **45**(9A): p. 7063–7079.
3. Ogawa, S., J. Tang, A. Yoshigoe, S. Ishidzuka, and Y. Takakuwa, Enhancement of $SiO_2/Si(001)$ interfacial oxidation induced by thermal strain during rapid thermal oxidation. *The Journal of Chemical Physics*, 2016. **145**(11): p. 114701.
4. Wang, W., K. Akiyama, W. Mizubayashi, T. Nabatame, H. Ota, and A. Toriumi, Effect of Al-diffusion-induced positive flatband voltage shift on the electrical characteristics of Al-incorporated high-k metal-oxide-semiconductor field-effective transistor. *Journal of Applied Physics*, 2009. **105**(6): p. 064108.
5. Wang, W., W. Mizubayashi, K. Akiyama, T. Nabatame, and A. Toriumi, Systematic investigation on anomalous positive V_{fb} shift in Al-incorporated high-k gate stacks. *Applied Physics Letters*, 2008. **92**(16): p. 162901–162903.
6. Iwamoto, K. et al., Experimental evidence for the flatband voltage shift of high-k metal-oxide-semiconductor devices due to the dipole formation at the high-k/SiO_2 interface. *Applied Physics Letters*, 2008. **92**(13): p. 132907–132903.

7. Kamimuta, Y. et al. Comprehensive study of V_{FB} shift in High-k CMOS - dipole formation, fermi-level pinning and oxygen vacancy effect in electron devices meeting, 2007. in IEDM 2007. IEEE International. 2007.

8. De, I., D. Johri, A. Srivastava, and C.M. Osburn, Impact of gate workfunction on device performance at the 50 nm technology node. *Solid-State Electronics*, 2000. **44**(6): p. 1077–1080.

9. Li, Z. et al., Flatband voltage shift of ruthenium gated stacks and its link with the formation of a thin ruthenium oxide layer at the ruthenium/dielectric interface. *Journal of Applied Physics*, 2007. **101**(3): p. 034503.

10. Hou, Y.T. et al., High performance tantalum carbide metal gate stacks for nMOSFET application. in *Electron Devices Meeting*, 2005. IEDM Technical Digest. IEEE International. 2005.

11. Wang, X., K. Han, W. Wang, X. Ma, D. Chen, J. Zhang, J. Du, Y. Xiong, and A. Huang, Comprehensive understanding of the effect of electric dipole at high-k/SiO2 interface on the flatband voltage shift in metal-oxide-semiconductor device. *Applied Physics Letters*, 2010. **97**(6): p. 062901.

12. Ragnarsson, L.A., N.A. Bojarczuk, J. Karasinski, and S. Guha, Hall mobility in hafnium oxide based MOSFETs: Charge effects. *Electron Device Letters, IEEE*, 2003. **24**(11): p. 689–691.

13. Choi, K., H.C. Wen, G. Bersuker, R. Harris, and B.H. Lee, Mechanism of flatband voltage roll-off studied with Al_2O_3 film deposited on terraced oxide. *Applied Physics Letters*, 2008. **93**(13): p. 133506.

14. Kraut, E.A., R.W. Grant, J.R. Waldrop, and S.P. Kowalczyk, Precise determination of the valence-band edge in X-ray photoemission spectra: Application to measurement of semiconductor interface potentials. *Physical Review Letters*, 1980. **44**(24): p. 1620–1623.

15. Zhu, L., K. Kita, T. Nishimura, K. Nagashio, S.K. Wang, and A. Toriumi, Observation of dipole layer formed at high-k dielectrics/SiO₂ interface with X-ray photoelectron spectroscopy. *Applied Physics Express*, 2011. **3**(6): p. 061501.

16. Zhu, L.Q., N. Barrett, P. Jegou, F. Martin, C. Leroux, E. Martinez, H. Grampeix, O. Renault, and A. Chabli, X-ray photoelectron spectroscopy and ultraviolet photoelectron spectroscopy investigation of Al-related dipole at the HfO_2/Si interface. *Journal of Applied Physics*, 2009. **105**(2): p. 024102–024107.

17. Liu, Z.Q., S.Y. Chiam, W.K. Chim, J.S. Pan, and C.M. Ng, Effects of thermal annealing on the band alignment of lanthanum aluminate on silicon investigated by x-ray photoelectron spectroscopy. *Journal of Applied Physics*, 2009. **106**(10): p. 103718–103719.

18. Duan, T.L., H.Y. Yu, L. Wu, Z.R. Wang, Y.L. Foo, and J.S. Pan, Investigation of HfO_2 high-k dielectrics electronic structure on SiO_2/Si substrate by x-ray photoelectron spectroscopy. *Applied Physics Letters*, 2011. **99**(1): p. 012902–0129023.

19. Afanas'ev, V.V. and A. Stesmans, Internal photoemission at interfaces of high-kappa insulators with semiconductors and metals. *Journal of Applied Physics*, 2007. **102**(8): p. 081301–081328.

20. Widicz, J., K. Kita, K. Tomida, T. Nishimura, and A. Toriumi, Internal photoemission over HfO_2 and $Hf(1-x)SixO_2$ high-k insulating barriers: Band offset and interfacial dipole characterization. *Japanese Journal of Applied Physics*, 2008. **47**(4): p. 2410–2414.

21. Kirsch, P.D. et al., Dipole model explaining high-k/metal gate field effect transistor threshold voltage tuning. *Applied Physics Letters*, 2008. **92**(9): p. 092901–092903.

22. Sivasubramani, P. et al., Dipole moment model explaining nFET V_t Tuning Utilizing La, Sc, Er, and Sr Doped HfSiON Dielectrics. in *VLSI Technology*, 2007 IEEE Symposium on 2007.

23. Kita, K. and A. Toriumi. Intrinsic origin of electric dipoles formed at high-k/SiO_2 interface. in *Electron Devices Meeting*, 2008. IEDM 2008. IEEE International 2008.

24. Kita, K. and A. Toriumi, Origin of electric dipoles formed at high-k/SiO_2 interface. *Applied Physics Letters*, 2009. **94**(13): p. 132902–132903.

25. Wang, X. et al., Physical origin of dipole formation at high-k/SiO_2 interface in metal-oxide-semiconductor device with high-k/metal gate structure. *Applied Physics Letters*, 2010. **96**(15): p. 152907–152903.

26. Heine, V., Theory of surface states. *Physical Review*, 1965. **138**(6A): p. A1689–A1696.

27. Parker, G.H., T.C. McGill, C.A. Mead, and D. Hoffman, Electric field dependence of GaAs Schottky barriers. *Solid-State Electronics*, 1968. **11**(2): p. 201–204.

28. Andrews, J.M. and M.P. Lepselter, Reverse current-voltage characteristics of metal-silicide Schottky diodes. *Solid-State Electronics*, 1970. **13**(7): p. 1011–1023.

29. Crowell, C.R., A simplified self-consistent model for image force and interface charge in Schottky barriers. *Journal of Vacuum Science and Technology*, 1974. **11**(6): p. 951–957.

30. Bardeen, J., Surface states and rectification at a metal semi-conductor contact. *Physical Review*, 1947. **71**(10): p. 717–727.

31. Cowley, A.M. and S.M. Sze, Surface states and barrier height of metal-semi-conductor systems. *Journal of Applied Physics*, 1965. **36**(10): p. 3212–3220.

32. Raymond T., Recent advances in Schottky barrier concepts. *Materials Science and Engineering: R: Reports*, 2001. **35**(1–3): p. 1–138.

33. Louie, S.G., J.R. Chelikowsky, and M.L. Cohen, Ionicity and the theory of Schottky barriers. *Physical Review B*, 1977. **15**(4): p. 2154–2162.

34. Tersoff, J., Theory of semiconductor heterojunctions: The role of quantum dipoles. *Physical Review B*, 1984. **30**(8): p. 4874–4877.

35. Robertson, J. and C.W. Chen, Schottky barrier heights of tantalum oxide, barium strontium titanate, lead titanate, and strontium bismuth tantalate. *Applied Physics Letters*, 1999. **74**(8): p. 1168–1170.

36. Robertson, J., *Band Offsets of Wide-Band-Gap Oxides and Implications for Future Electronic Devices*. (Raleigh, NC, AVS, 2000).
37. Monch, W., Slope parameters of the barrier heights of metal-organic contacts. *Applied Physics Letters*, 2006. **88**(11): p. 112116–112113.
38. Fonseca, L.R.C., D. Liu, and J. Robertson, p-type Fermi level pinning at a Si:Al$_2$O$_3$ model interface. *Applied Physics Letters*, 2008. **93**(12): p. 122905–122903.
39. Monch, W., On the electric-dipole contribution to the valence-band offsets in semiconductor-oxide heterostructures. *Applied Physics Letters*, 2007. **91**(4): p. 042117–042113.
40. Wang, X., K. Han, W. Wang, X. Ma, J. Xiang, D. Chen, and J. Zhang, Electric dipole at high-k/SiO$_2$ interface and physical origin by dielectric contact induced gap states. *Japanese Journal of Applied Physics*, 2011. **50**(10): p. 10PF02.
41. Lee, B.H., J. Oh, H.H. Tseng, R. Jammy, and H. Huff, Gate stack technology for nanoscale devices. *Materials Today*, 2006. **9**(6): p. 32–40.
42. Song, S.C. et al., Mechanism of V$_{fb}$ roll-off with high work function metal gate and low temperature oxygen incorporation to achieve PMOS band edge work function. in *Electron Devices Meeting*, 2007. IEDM 2007. IEEE International. 2007.
43. Akiyama, K., W. Wang, W. Mizubayashi, M. Ikeda, H. Ota, T. Nabatame, and A. Toriumi, V$_{FB}$ roll-off in HfO$_2$ gate stack after high temperature annealing process - a crucial role of out-diffused oxygen from HfO$_2$ to Si. in *VLSI Technology*, 2007 IEEE Symposium on. 2007.
44. Akiyama, K., W. Wang, W. Mizubayashi, M. Ikeda, H. Ota, T. Nabatame, and A. Toriumi, Roles of oxygen vacancy in HfO$_2$/ultra-thin SiO$_2$ gate stacks - comprehensive understanding of VFB roll-off. in *VLSI Technology*, 2008 Symposium on. 2008.
45. Bersuker, G. et al., Origin of the flatband-voltage roll-off phenomenon in metal/high-k gate stacks. *IEEE Transactions on Electron Devices*, 2010. **57**(9): p. 2047–2056.
46. Choi, K., H.C. Wen, G. Bersuker, R. Harris, and B.H. Lee, Mechanism of flatband voltage roll-off studied with Al$_2$O$_3$ film deposited on terraced oxide. *Applied Physics Letters*, 2008. **93**(13): p. 133506.
47. Zheng, X.H., A.P. Huang, Z.S. Xiao, Z.C. Yang, M. Wang, X.W. Zhang, W.W. Wang, and P.K. Chu, Origin of flat-band voltage sharp roll-off in metal gate/high-k/ultrathin- SiO$_2$/Si p-channel metal-oxide-semiconductor stacks. *Applied Physics Letters*, 2010. **97**(13): p. 132908.

CHAPTER 2

MOS Processes

2.1 MOS CAPACITOR PREPARATION PROCESS

MOS capacitor refers to a device that uses a metal–oxide–semiconductor structure as a capacitor. The specific structure is shown in Figure 2.1. The device has two electrodes, namely, a metal and a semiconductor, which are separated by a dielectric. The metal electrode is the gate, and the semi-conductor end is the back gate or body. The insulating oxide layer between them is called the gate dielectric. The MOS capacitor in Figure 2.1 can be regarded as a parallel plate capacitor. When a certain voltage is applied to the two poles of the capacitor, the metal gate and the semiconductor substrate will be charged. They carry the same amount of charge, but the polarity of the charge is opposite. But unlike simple parallel plate capaci-tors, the charge distribution in the metal gate and the semiconductor sub-strate is different. MOS capacitors are widely used in integrated circuits.

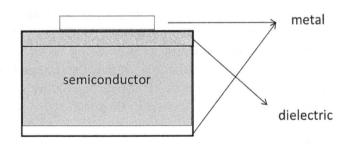

FIGURE 2.1   MOS capacitor structure.

DOI: 10.1201/9781003216285-3

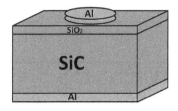

FIGURE 2.2   A schematic diagram of a SiC MOS capacitor.

The following takes SiC MOS capacitors as an example to illustrate the general preparation process of MOS capacitors in detail.

Figure 2.2 is a schematic diagram of a SiC MOS capacitor. The process of turning a wafer into an MOS capacitor includes a series of operations, such as slicing, cleaning, dielectric formation, metal masking, metal evaporation to form a gate and back electrode formation.

2.1.1 Slicing

Slicing refers to the process of cutting a large wafer into smaller wafers with a certain area. The surface of the wafer easily absorbs the dust particles in the air, so slicing is generally carried out in an ultraclean room. Slicing includes manual slicing and slicer slicing. Common slicers include blade slicers and more advanced laser slicers. Blade slicing is a mechanical form. The diamond knife used to divide the wafer will apply pressure on the wafer, causing the edge of the chip to crack, which can even damage the chip and cause a decline in yield. In addition, small cracks will extend, which greatly affects the subsequent reliability and service life of the chip. Laser slicing is a more advanced slicing process that uses high-energy laser beams to irradiate the wafer to fuse the irradiated area. The processing is noncontact, and there is no mechanical pressure on the wafer itself, so that the wafer will not be damaged. In the case of limited experimental conditions, manual slicing can be performed. Manual slicing has the same mechanism as blade slicing, which is to use a diamond pen to cut the wafer. Manual slicing will have the same problems as blade slicing, which is difficult to operate. Manual slicing requires a diamond pen, a graph paper, a protective film and a ruler (Figure 2.3).

2.1.2 Cleaning

The cleaning process is important in the semiconductor process, and the cleaning process directly affects the performance and reliability of the wafer and the devices made from the wafer. The cleaning process includes

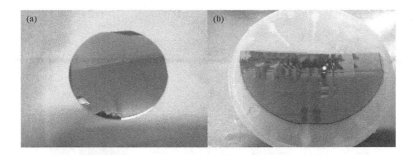

FIGURE 2.3   SiC wafers (a) before and (b) after slicing.

dry cleaning and wet cleaning. In simple terms, dry cleaning is the use of gases to clean the wafers, such as thermal chemical gases and plasma reaction gases, while wet cleaning is the use of solutions to clean the wafers. Wet cleaning is the more common cleaning method. Common pollutants include particles, metals, organic matter and natural oxide layer. Choose the appropriate solution according to the type of pollutant. The RCA cleaning method is the first wet cleaning process for wafer cleaning. The cleaning method relies on solvents, acids, surfactants and water to spray, purify, oxidize, etch and dissolve without destroying the surface characteristics of the wafer. Contamination on the surface of the wafer can be organic matter or metal ion contamination. Wash thoroughly in deionized water after each use of chemicals. The following briefly introduces the cleaning fluid used by RCA and its function: APM (ammonia hydroxide-hydrogen peroxide water mixture) is usually called SC1 cleaning solution, its formula is $NH_4OH : H_2O_2 : H_2O = 1{:}1{:}5{\sim}1{:}2{:}7$, and it is used to remove surface particles by oxidation and micro-etching. HPM (hydrochloric/peroxide mixture) is usually called SC2 cleaning solution, its formula is $HCl : H_2O_2 : H_2O = 1{:}1{:}6{\sim}1{:}2{:}8$, and it can dissolve alkali metal ions and hydroxides of aluminum, iron and magnesium. In SPM (sulfuric peroxide mixture) cleaning solution, the volume ratio of sulfuric acid to water is 1:3. It is a typical cleaning solution used to remove organic pollutants. DHF cleaning solution is diluted hydrofluoric acid, which can remove the natural oxide film on the surface of the silicon wafer. This cleaning method was used for reference by many subsequent cleaning processes.

According to the requirements of this SiC MOS capacitor preparation experiment, the SiC wafers are washed with acetone, ethanol and deionized water successively. In order to improve the cleaning effect, ultrasonic treatment can be carried out. Acetone can remove organics on SiC chips,

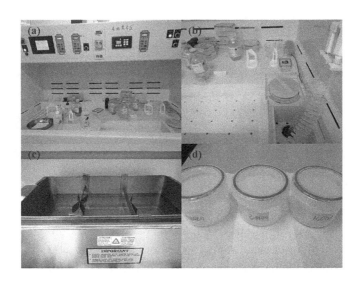

FIGURE 2.4   (a) Manual wet cleaning tool; (b) wet cleaning tank; (c) ultrasonic cleaning machine; (d) solution boxes.

ethanol can remove acetone residue, and deionized water can remove ethanol. Acetone, ethanol and deionized water are placed in their own containers, and the containers cannot be used in series (Figure 2.4).

The final step in the cleaning process is to dry the wafers. To dry a wafer with a nitrogen gun, the nitrogen flow should not be perpendicular to the surface of the wafer. One end of the wafer can be placed on the dust-free paper, and the other end is tilted. The nitrogen flow is parallel to the wafer surface to dry the wafer surface. Pay attention to the direction of nitrogen flow from the chip to the tweezers; otherwise, the dirty water on the tweezers may flow to the chip (Figure 2.5).

2.1.3 Dielectric Formation

The role of dielectric is to isolate metal and semiconductor. In MOS capacitors, dielectric refers to the oxide layer. Taking Si MOS capacitors and SiC MOS capacitors as examples, the oxide layer is usually SiO_2 grown on the substrate by oxidation or deposited SiO_2. In this SiC MOS capacitor preparation experiment, two processing methods for forming SiO_2 layer are introduced. The first method is to thermally oxidize the SiC wafer to form an oxide layer. In the oxidation furnace, the high-temperature environment and oxygen atmosphere are required for the formation of a dense oxide layer. In the thermal oxidation process, the airtightness of

FIGURE 2.5   Blowing SiC wafer with nitrogen gun.

the oxidation furnace is a problem worthy of attention. At the same time, it should be noted that it is strictly forbidden to put metal electrodes or wafers contaminated by metal into the high-temperature oxidation furnace. Metal ions will seriously affect the quality of the oxide layer. The second method is microwave plasma oxidation of the SiC wafer to form an oxide layer. Plasma oxygen has higher energy than gaseous oxygen. In a microwave plasma generator, the electric field excites the gaseous oxygen into plasma oxygen, and the plasma oxygen reacts with the SiC on the substrate surface to form a SiO_2 layer (Figure 2.6).

FIGURE 2.6   (a, b) High-temperature annealing furnace; (c) microwave plasma facilities.

2.1.4 Metal Evaporation to Form Electrodes

After the oxide layer is grown on the wafer surface, the metal is deposited on the oxide layer to form a metal gate, and the metal oxide–semiconductor structure is completed. In the manufacturing process of discrete components and simple circuits, vacuum evaporation of metal is the most common metal deposition method. Figure 2.7a is an image of a conventional vacuum evaporation system. The vacuum pump and the quartz bell are the two main parts of the vacuum system. The vacuum environment can prevent the wafer from being contaminated and the metal vapor from being oxidized. At the same time, it can also make the thickness of the metal film evaporated on the wafer more uniform. In related experiments or tests of MOS capacitors, the area and shape of

FIGURE 2.7 (a) Vacuum evaporation table; (b) metal mask; (c) metal gate and back electrode of SiC MOS capacitor.

the metal gate have certain requirements. This requires the use of a metal mask when evaporating the metal gate. Evaporating the metal is to print the shape of the mask on the wafer. The metal mask can be designed by CAD drawing software and made by a specialized manufacturer. To prepare for evaporation, a nitrogen gun blows the wafer surface to prevent particles from affecting the reliability of the MOS capacitor oxide layer. Then, the wafer is attached to the appropriate position on the metal mask. In this step, the oxide side and the nonoxidized side of the wafer are confirmed to prevent the order from being reversed and lead to erroneous experiments. At the same time, attention should be paid to avoid the relative sliding between the wafer and the metal mask to avoid damaging the oxide layer. Unfixed wafers may cause ghosting of the metal gate during the evaporation process. After confirming that the wafer is fixed on the metal mask, the metal mask and the wafer can be placed in the vacuum chamber. After the metal gate is formed, the wafer is taken out, and the metal is evaporated on the unprocessed side to form the back electrode. The semiconductor pole is either of the two poles of the MOS capacitor, but in the electrical test process, the semiconductor pole is not in close contact with the metal holder of the probe station, which seriously interferes with the test. Evaporating the metal back electrode can solve this problem. Metal–semiconductor contacts may form Schottky contacts or ohmic contacts. To make the resistance of the contact surface much smaller than the resistance of the semiconductor itself, and the contact surface has as little voltage drop as possible, it is necessary to form a good ohmic contact. To form a good ohmic contact, two conditions must be met: the metal and the semiconductor should have a low barrier height and the semiconductor should have a high impurity concentration. There is a low barrier height between the metal Al and SiC, and the high impurity concentration of the semiconductor can be achieved by using a diamond pen to rough the back. The diamond pen roughens the back of the SiC sheet and introduces a large number of defects on the back surface. Attention should be paid to the operation on the back of the roughing treatment: (1) the stroke direction of the diamond pen can only be in one direction, and it should not be stroked back and forth. This is conducive to cleaning the small SiC debris with a nitrogen gun; (2) the scratches should be evenly distributed on the back as much as possible.

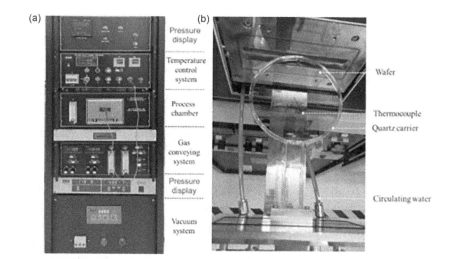

FIGURE 2.8 (a) Photo of rapid heat treatment equipment; (b) schematic illustration of quartz carrier.

2.2 OXIDATION PROCESS AND KINETICS

2.2.1 Thermal Processing (RTP) and Plasma Oxidation Systems

Based on the traditional thermal oxidation and the novel plasma oxidation, Figure 2.8 describes two systems with the example of rapid thermal oxidation and microwave plasma oxidation, including the configuration of the equipment and the operation method.

2.2.1.1 Thermal Processing (RTP) Systems

The basic equipment used for thermal oxidation includes horizontal furnaces, vertical furnaces and rapid thermal processing (RTP) furnaces. The first two types are often used to process a large number of silicon wafers at the same time. Horizontal furnaces were widely used in the early semiconductor industry. They were named so because the quartz tubes for placing and heating silicon wafers in them were in a horizontal position. But it has been gradually replaced by vertical furnaces because vertical furnaces are more automated, have better control temperature and uniformity, and reduce particle contamination. RTP is a small rapid heating system, usually a single-chip operation. It has the characteristics of fast and local heating. A typical RTP device can achieve a temperature rise and fall with the rate of tens of degrees per second.

In order to understand the basic thermal oxidation process of silicon, we take the RTP system as an example for illustrating. The RTP system is mainly divided into four parts: process chamber, gas system, temperature control system and control system.

- Process chamber: The process chamber is a place for heating silicon wafers. It is composed of a quartz carrier, a quartz bell jar and heating components. Specifically, the quartz carrier is a supporting structure for placing silicon wafers. The inner part of the quartz bell jar is a reaction chamber, and the outer part is a heating component, which is composed of a heating resistance wire and a heating tube sleeve. The process chamber is connected with the gas system, and the temperature control system is around the outside.

- Gas system: The gas system includes the reaction gas conveying system supplied to the process chamber and the vacuum system for gas removal. In order to realize the reaction atmosphere in the process chamber, the gas transportation system includes a gas source and a corresponding pipeline. The vacuum system is used to remove the useless gas in the chamber before and after the reaction. It is also used during the reaction with the gas delivery system to maintain the required pressure. The vacuum system realizes the vacuum by mechanical pump and turbo molecular pump. The mechanical pump achieves the pressure $< 3 \times 10^{-1}$ Pa in the chamber, and the turbine molecular pump achieves the pressure $< 3 \times 10^{-3}$ Pa.

- Temperature control system: The temperature control system is used to adjust the reaction temperature. The core part is the sensor, usually a thermocouple, which provides the corresponding millivolt signal to the furnace controller while detecting the temperature. In addition to thermocouples, the temperature control system also includes the cooling system consisting of a fan and circulating water.

- Control system: The control system is used to set the temperature and start the reaction. Specifically, the corresponding experimental temperature curve is set on the computer before the experiment, and the temperature is automatically adjusted by the feedback signal of thermocouple in the reaction process.

The specific operation steps of the RTP system are as follows:

1. Turn on the main power supply of the equipment, then turn on the power supply of the temperature control system module and the gas system module, respectively.

2. Open the valve of the corresponding gas transmission pipeline, inject nitrogen into the process chamber, maintain the process chamber at 1 atm, and then open the chamber. Check and adjust the position of the thermocouple (it is better to be close to but not in contact with the back of the wafer).

3. Close the chamber and stop the injection of nitrogen. Turn on the mechanical pump and vacuum gauge. After the pressure reaches $< 3 \times 10^{-1}$ Pa, turn on the turbomolecular pump and wait for the pressure to reach $< 3 \times 10^{-3}$ Pa.

4. Press "Run" button in the device control panel, then set the pre-bake temperature curve on the computer. Tap on the "Start" button on the computer, and observe the temperature change curve.

5. After the end of pre-bake, lift up the "Run" button in the equipment, turn off the mechanical pump and vacuum gauge, and wait for the equipment to cool down to below 80°C.

6. Once again, nitrogen is injected into the process chamber to maintain the process chamber at 1 atm, and then the chamber is opened. Put the sample on the quartz carrier, and check the position of the thermocouple (it is better to be close to but not in contact with the back of the sample). Repeat step 3.

7. Turn on the turbomolecular pump, and turn down the mechanical pump. Open the corresponding gas pipeline valve, and inject oxygen into the process chamber until the needed pressure is reached.

8. Press "Run" button in the device control panel, then set the experimental temperature curve on the computer. Tap on the "Start" button on the computer, and observe the temperature change curve.

9. After oxidation, lift up the "Run" button in the equipment, and turn up the mechanical pump and close the gas valve. Wait for the

equipment to cool down to below 80°C. Close the mechanical pump and vacuum gauge of the oxygen delivery pipeline.

10. Inject nitrogen into the process chamber to maintain the process chamber at 1 atm, then open the cavity and take out the sample.

11. In order to keep the chamber clean, vacuum treatment is needed after the experiment. After vacuum pumping, turn off all power supply and gas valves of the equipment (Figure 2.9).

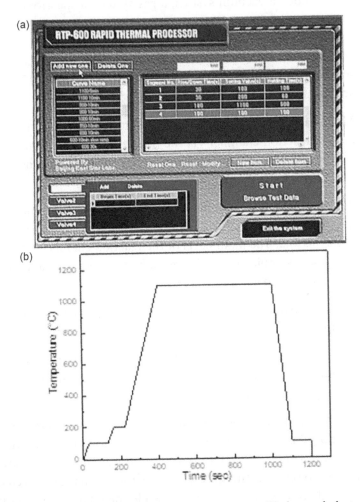

FIGURE 2.9   (a) Program to set the temperature curve; (b) the pre-bake temperature curve.

2.2.1.2 Plasma Oxidation Systems

In order to understand the plasma oxidation process, we take the microwave plasma oxidation system as an example for illustrating. Compared with the other plasma systems such as capacitance coupling plasma and inductive coupling plasma, microwave plasma has several intrinsic advantages such as high reactivity, low collision kinetic energy and no contamination of the electrode material. The microwave plasma oxidation system mainly consists of a microwave source, a microwave transmission part, a reaction chamber, a vacuum system, a gas system, a spectrometer and a temperature monitoring system.

- Microwave source: The self-oscillation of magnetron, as a power source, can output microwave with a frequency of 2.45 GHz and adjustable power in the range of 300~2000 W.

- Microwave transmission part: The microwave is transmitted to the plasma reaction chamber via the transmission part. It mainly consists of a rectangular waveguide, a tuning adapter and a short-circuit piston.

- Reaction chamber: The microwave energy is directly coupled into the reaction chamber filled with pure oxygen to complete excitation of the plasma. The main parts of the reaction chamber are a quartz tube and a sample holder. A plasma hemisphere with a diameter of 2.5 cm is located at the center of the chamber.

- Vacuum system: The reaction chamber is evacuated by mechanical pumps to < 0.01 Pa.

- Gas system: Both the gas system and the vacuum system are connected to the quartz tube. The gas system includes 16-oxygen, 18-oxygen and nitrogen.

- Spectrometer: The composition and density of the plasma in the reaction chamber are obtained by real-time monitoring of the emission wavelength and intensity by the optical emission spectra. The fiber probe collects the optical signal of the plasma reaction chamber through an observation window beside the quartz tube and transmits it to the spectrometer. The optical signal is converted into an electrical signal and then presented on a computer.

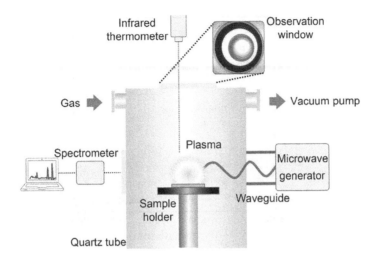

FIGURE 2.10   Schematic of microwave plasma oxidation equipment.

- Temperature monitoring system: The infrared thermometer is used to detect the real-time temperature of the sample surface, and the thermocouple at the bottom of the sample holder is used to detect the real-time temperature of the sample holder (Figure 2.10).

The specific operation steps of the plasma oxidation system are as follows:

1. Turn on the main power supply of the equipment. Open the valve of the corresponding gas transmission pipeline, inject nitrogen into the process chamber, maintain the process chamber at 1 atm, and then open the chamber. Put the sample on the sample holder.

2. Close the chamber, and stop nitrogen injection. Turn on the mechanical pump and vacuum gauge, and wait for the pressure to reach $< 3 \times 10^{-1}$ Pa.

3. The chamber is filled with 0.5 kPa of pure oxygen. Turn on the power supply of the microwave source, and set the input power of 500 W to excite the plasma.

4. After the plasma is formed, the power and pressure are adjusted to the conditions required by the experiment. At the same time, the surface temperature of the sample and the emission spectrum of the plasma can be monitored in real time.

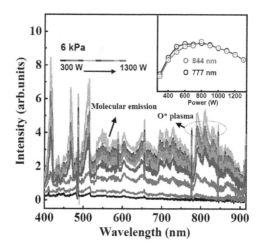

FIGURE 2.11   Optical emission spectra of oxygen plasma at various input power.

5. After the end of process, inject nitrogen into the process chamber to maintain the process chamber at 1 atm, then open the cavity and take out the sample.

6. In order to keep the chamber clean, vacuum treatment is needed after the experiment. After vacuum pumping, turn off all power supply and gas valves of the equipment (Figure 2.11).

Example 2.1: Si Thermal Oxidation Kinetics

In integrated circuits, silicon dioxide has many uses, such as gate dielectric of MOSFETs, mask for diffusion or implantation of impurities, isolation of devices and passivation protection of circuits, etc. The silicon dioxide can be obtained by thermal oxidation or deposition. In thermal oxidation, high-purity oxygen or water vapor is supplied from the ambient to Si/SiO_2 interface in a heated environment and reacts with the silicon substrate, thereby obtaining a thermally grown oxide layer on the silicon wafer. The deposited oxide layer can be supplied with high-purity oxygen and silicon source from the outside to make them react in the chamber to form a film on the surface of the silicon wafer, such as plasma-enhanced chemical vapor deposition (PECVD). In this section, we only focus on the thermal oxidation process of Si.

For thermal growth, the growth of silicon dioxide consumes silicon. The thickness of silicon consumed accounts for 46% of the total thickness of the oxide, i.e., for every 100-nm-thick oxide, 46-nm-thick silicon is consumed. The thermal growth of Si is divided into dry thermal oxidation and wet thermal oxidation. If the oxidant is high-purity oxygen, the chemical reaction formula is as follows:

$$Si + O_2(g) \rightarrow SiO_2 \tag{2.1}$$

In the oxidation, the reaction temperature, gas pressure, reaction time and other factors will affect the thickness and quality of the oxide layer. Generally, the temperature of the thermally grown oxide layer on silicon is between 750°C and 1100°C. If the oxidant is water vapor, the chemical reaction formula is as follows:

$$Si + 2H_2O(g) \rightarrow SiO_2 + 2H_2(g) \tag{2.2}$$

Compared with dry thermal oxidation, wet thermal oxidation has a faster growth rate, because water vapor diffuses faster in silica and has a higher solubility than oxygen. However, the hydrogen molecules generated by the reaction are bound in the silicon dioxide layer, which reduces the density of the oxide layer.

The growth model of oxide on Si is a linear–parabolic model developed by Deal and Grove, which can accurately describe the oxide growth in a wide range of thickness, but is limited in the thinner oxide thickness range. The model assumes that the oxidation process is caused by the transport and interaction of oxidant species:

1. The oxidant is transported from the bulk of the oxidizing gas to the outer surface of the oxide.
2. The oxidant diffuses across the oxide film.
3. The oxidant reaches the interface and reacts with silicon to form a new layer of SiO_2.

The three stages of the silicon oxidation process are shown in Figure 2.12. Taking into account the reaction and diffusion that

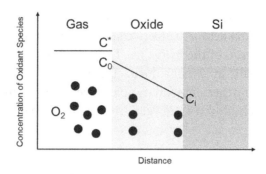

FIGURE 2.12   Schematic illustration of Deal–Grove model.

occur at the two boundaries of the oxide layer, the model derives the relationship:

$$x^2 + Ax = B(t + \tau) \tag{2.3}$$

where x is the thickness of the oxide layer, t is the oxidation time, B is the parabolic rate constant, B/A is the linear rate constant, and τ corresponds to a shift in the time coordinate which corrects for the presence of the initial oxide layer. According to Equation (2.3), it can be found that the initial stage of silicon dioxide growth is a linear stage, which is described by a linear equation:

$$x = (B / A)t \tag{2.4}$$

In the linear region, the thickness of the oxide layer changes linearly with time. The limiting factor at this stage is the reaction on the Si/SiO$_2$ interface. The second stage of oxidation is the parabolic stage, which usually starts after the oxide thickness is about 15 nm. The formula used to describe the parabolic region is:

$$x = (Bt)^{1/2} \tag{2.5}$$

The oxide growth in this region is much slower than in the linear phase. When the oxide layer becomes thicker, the oxygen has to pass through the thick oxide layer to the Si/SiO$_2$ interface before reacting with Si. Therefore, the oxidation reaction is limited by the diffusion rate of oxygen through the oxide layer.

According to the Deal–Grove model, the key to establishing the kinetics of Si oxidation is to determine the values of each parameter. By fitting the data, the following procedure is employed to obtain the parameters:

1. A detailed plot of oxide thickness vs oxidation time is made to determine what initial oxide thickness x_i the data extrapolated to at $t=0$. In the case of wet-oxygen oxidation, x_i was found to be zero at all temperatures. The plot for dry oxygen appears to pass through the thickness axis at $x_0 = 25$ nm.
2. The oxide thickness x_0 was plotted vs the quantity t/x_0. As can be seen from Equation (2.3), if that relationship holds and $\tau=0$, such a plot should yield straight lines with the intercept equal to $-A$, and the slope equal to B.

The pressure of rate constants is discussed. In the Deal–Grove model, B and B/A can be expressed by:

$$B = 2D_{\text{eff}} \frac{C^*}{N} \tag{2.6}$$

$$\frac{B}{A} = \frac{kh}{k+h} \frac{C^*}{N} \approx k \frac{C^*}{N} \tag{2.7}$$

where C^* is the equilibrium concentration of the oxidant in SiO_2, N is the number of oxidants incorporated into a unit volume of the SiO_2, D_{eff} is the effective diffusion coefficient, and h is a gas-phase transport coefficient. According to Equation (2.6), the parabolic rate constant B should be proportional to C^*, which in turn should be proportional to the partial pressure of the oxidizing species in the gas if Henry's law is obeyed. Conversely, the coefficient A should be independent of the partial pressure.

The temperature of rate constants is discussed. The logarithm of the parabolic rate constant B and B/A is plotted against the reciprocal of the absolute temperature. Good straight lines show the temperature dependence of B, which is substantially the same as that of D_{eff}. The activation energy of parameters can be extracted based on slope. The temperature dependence of the linear rate constant B/A can also be obtained in the same way.

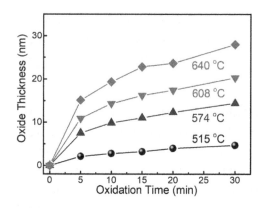

FIGURE 2.13   Time dependence of oxide thickness at various temperatures.

Example 2.2: SiC Plasma Oxidation Kinetics

In this section, we introduce how to study kinetics by taking SiC plasma oxidation as an example:

1. Study plasma oxidation rate under variable conditions, such as pressure, power, temperature, ground-state oxygen atom concentration, etc. The quasi-parabolic thickness-time curves for various temperatures and concentrations of ground-state oxygen (Figure 2.13).

2. Isotope tracing technique is used to study the diffusion kinetics. Specifically, two groups of samples are prepared by employing the ^{16}O and ^{18}O isotopes as tracers to study the transport of oxygen during plasma oxidation. The oxide growths are performed under plasma oxygen ambient at a pressure of 5 kPa, microwave power of 500 W and temperature of 630°C. For the first group (^{16}O-^{18}O), two samples are first oxidized in ^{16}O-plasma ambient for 5 min. After this process, 13.7-nm-thick $Si^{16}O_2$ layers are grown on both samples (denoted as $Si^{16}O_2$ (13.7 nm)/SiC). Then, one $Si^{16}O_2$ (13.7 nm)/SiC sample is reoxidized in ^{18}O-plasma ambient for 4 min, by which is obtained the $Si^{16,18}O_2$ (16.7 nm)/SiC. After all oxidation, the analysis of the depth distributions of ^{18}O, ^{16}O, ^{28}Si, and ^{12}C in the oxide is characterized by secondary ion mass spectrometry analysis. After the first step of ^{16}O-plasma oxidation, ^{16}O is found to distribute uniformly throughout the oxide. After the reoxidation in ^{18}O

plasma ambient, the external ^{18}O (external oxygen called "O_{ex}") builds up and broadens at the surface and interface in the SiO_2, while the lattice ^{16}O (lattice oxygen called "O_{la}") exhibits an arch distribution. In order to eliminate the uncertainty of the SIMS results caused by matrix and knock-on effects, the second group (^{18}O-^{16}O) is oxidized by a reverse two-step process at the same condition. Similar results strongly indicate that changes in oxygen distribution are truly phenomena of SiC plasma oxidation, rather than SIMS artifacts. After reoxidation, the intensity of the lattice oxygen is significantly reduced, indicating that oxygen exchange is accompanied by the outdiffusion of lattice oxygen into the ambient during the microwave plasma oxidation (Figures 2.14 and 2.15).

3. Quantitaive discussion on the oxygen diffusion in terms of the diffusion coefficient is executed based on the depth profiles of oxygen. If a constant ^{18}O concentration is maintained (i.e.,

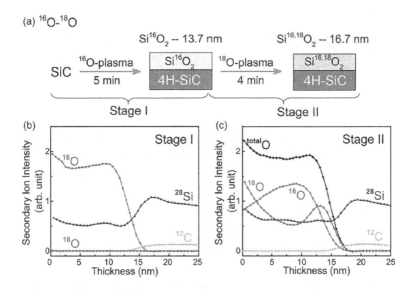

FIGURE 2.14 Schematics of (a) SiC oxidized in ^{16}O-plasma ambient and reoxidized in ^{18}O-plasma ambient. The temperature is 630°C. SIMS profiles of ^{16}O, ^{18}O, ^{28}Si, ^{12}C of the samples: (b) $Si^{16}O_2$ (13.7 nm)/SiC; (c) $Si^{16,18}O_2$ (16.7 nm)/SiC. (Reproduced from N. N. You et al., Vacuum 182, 109762 (2020), with the permission of Elsevier Publishing.)

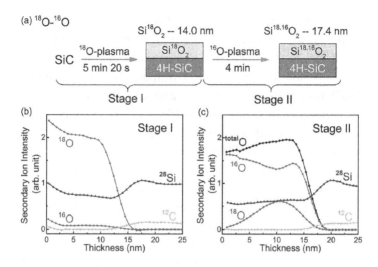

FIGURE 2.15 Schematics of (a) SiC oxidized in ^{18}O-plasma ambient and reoxidized in ^{16}O-plasma ambient. The temperature is 630°C. SIMS profiles of ^{16}O, ^{18}O, ^{28}Si, ^{12}C of the samples: (b) Si^{18}O$_2$ (14.0 nm)/SiC; (c) Si^{18,16}O$_2$ (17.4 nm)/SiC. (Reproduced from N. N. You et al., Vacuum 182, 109,762 (2020), with the permission of Elsevier Publishing.)

$C(0, t) = \text{constant} = C_0$ at the surface and adjacent region in SiO$_2$, the ^{18}O concentration (C) and the depth (x) can be given as

$$C(x,t) = C_s \text{erfc}(\frac{x}{2\sqrt{D_s t}}) \qquad (2.8)$$

where D_s is the diffusion coefficient of oxygen near the surface, t is the oxidation time, and C_s is the ^{18}O concentration at the surface. If only a finite source of oxygen, Q_0, can be used for this exchange process during the anneal, the oxygen concentration profile is given as

$$C(x,t) = \left(Q_0 / \sqrt{\pi D T}\right) \exp\left(-x^2 / 4Dt\right) \qquad (2.9)$$

and C_{int}, the measured ^{18}O concentration at the interface, will equal to $Q_0/\sqrt{\pi D T}$. In fact, the diffusivity can be estimated quickly by the following method. Assuming the diffusion depth x_0 of 270 nm and $x_0 = \sqrt{D t_0}$, where $t_0 = 21600$ s is annealing time, the self-diffusion, where coefficient D^* of the ^{18}O in the SiO_2 is estimated to be$=3.4\times10^{-14}cm^2/s$ at 1360°C. The value is the upper limit, as the diffusion depth may refer to $2\sqrt{Dt_0}$ or $4\sqrt{Dt_0}$.

4. The quantitative model for SiC plasma oxidation is established, and temperature and oxidant concentration dependent on oxidation behaviors are systematically discussed. We applied Massoud's empirical relation for SiC plasma oxidation, studying oxide thickness vs time for various temperatures and deducing the growth rate parameters. Massoud's empirical relation includes the exponential term based on the Deal–Grove model [4–6].

$$\frac{dx}{dt} = \frac{B}{A+2x} + C\exp\left(-\frac{x}{L}\right) \qquad (2.10)$$

where B and B/A are the parabolic rate constant and the linear rate constant, respectively, and C and L refer to the pre-exponential constant and the characteristic length during growth rate enhancement, respectively.

First, we need to make some reasonable assumptions as prerequisites before fitting, because it is necessary to limit the initial value of the parameters when the fitting data is not enough. For example, the growth rate B describes the diffusivity in SiO_2; therefore, the nature of the oxide determines the value of B that is identical for all surface orientations. Goto et al. proposed that growth rate B is related to the diffusion of oxidants in SiO_2, but not to surface orientation [7]. Thus, B can be given by oxygen self-diffusivity in oxide, just like Si oxidation. Since 1.5 times of oxygen consumption is in SiC oxidation compared with Si oxidation, [8] the B value for SiC can be expressed by:

$$B = \frac{2D_O^{SD}}{1.5} = 4.27\times10^{-8}\times\exp\left(-\frac{1.64eV}{k_BT}\right)cm^2/s \qquad (2.11)$$

According to Massoud's empirical relation and the assumptions, the plots of oxide time vs oxide thickness can be fitted. The relationship of parameters with temperature can be obtained, and the activation energy (E_a) of the growth rate coefficients can be extracted from Arrhenius plots.

In the same way, we can discuss other conditions, such as pressure, power, ground-state oxygen atom concentration, etc.

5. A plausible model to explain the kinetics of the plasma oxidation of SiC needs to be proposed based on the oxygen transfer. For example, after reoxidation, part of the external oxygen exists at the interface, which indicates that part of the external oxygen diffuses through the SiO_2 and reacts with SiC to form new SiO_2. Oxygen exchange between the SiO_2 and the ambient is responsible for the external oxygen distribution in the surface and adjacent region. We infer the exchange of the external oxygen with lattice oxygen occurs possibly in terms of SiO or O_2.

Example 2.3: Ge Active Oxidation

The reaction of oxygen with surfaces of semiconductor has attracted considerable interest, primarily due to its crucial role in the processing of microelectronic devices. For silicon, at sufficiently elevated temperatures, exposure of Si surfaces to O_2 or O will lead to the formation of volatile SiO. This is called active oxidation as opposed to passive oxidation in which a stable SiO_2 film is formed. For Ge, unlike SiO_2/Si, where the desorption of SiO is limited to only several monolayers (MLs), the desorption of GeO from GeO_2/Ge occurs severely even in very thick cases. Therefore, the oxidation behavior of Ge should be studied by taking into consideration the desorption of GeO. Before discussing the oxidation of Ge, we have to clarify the definition of active and passive oxidation, respectively. In this subsection, active oxidation refers to the oxidation with the consumption of semiconductor atoms. And passive oxidation is the oxidation without the consumption of semiconductor atoms.

For Ge, we have

$$\mathbf{Ge(s) + O_2 = GeO_2 \ (s) \ (Passive)} \tag{2.13}$$

$$\mathbf{Ge(s) + \frac{1}{2}O_2 = GeO \ (g) \ (Passive)} \tag{2.14}$$

In this subsection, a direct evidence of the active oxidation of Ge will be given through an atomic force microscope (AFM) observation of the Ge annealed under low pO_2. Actually, the low pO_2 condition is quite common in the semiconductor process flow. As previously demonstrated, the GeO desorption is mainly studied under the ultra-high vacuum (UHV), where the oxygen partial pressure is estimated to be below 10^{-10} Pa, and the role of the residual oxygen is negligibly small enough to be investigated. However, actual thermal processes such as dopant activation and forming gas annealing are difficult to carry out under UHV conditions. Therefore, instead of the UHV conditions, N_2 ambient annealing is usually employed to build an oxygen-free environment for thermal treatment. However, it is worth pointing out that, for example, for 1 atm N_2 ambient annealing, we can only achieve a relatively oxygen-free environment with the residual oxygen partial pressure generally around 0.1 Pa. This is much higher than that in the UHV conditions. Thus, we consider that the oxygen effect might be quite different from those in the UHV cases. Although the oxidation or oxygen adsorption on a semiconductor surface has been widely investigated, direct observation results are still very rare.

To investigate oxidation on a Ge substrate, stable oxide fins (some were 110 nm sputtered Y_2O_3 and some were 160-nm-thick spin-on-glass (SOG) SiO_2) with the size of 1.5 μm × 5 mm × 160 nm (width × length × thickness) were fabricated by using a photolithography technique. The distance between two GeO_2 fins was 1.5 μm, and the distance between two SiO_2 fins was 7.5 μm. The line-patterned structure is schematically shown in Figure 2.16a. And an AFM image of the initial surface is shown in Figure 2.16b. By using this structure, we could know the level of the initial Ge surface, because no matter what treatment we used, the Ge surface underneath the SiO_2 capping will be kept.

Figure 2.17 shows the cross-sectional profiles of as-received sample and those annealed at 600°C and oxygen partial pressure 0.1 Pa for 30, 90 and 150 min, respectively. Since the SOG fins are thick enough, it is quite reasonable to believe that the reaction between oxygen and Ge substrate only occurs at the places without SOG covering. The Ge substrate surface without SOG fins was clearly consumed as the annealing time increased at low oxygen partial pressure, suggesting that the Ge substrate is "etched" during the annealing treatment.

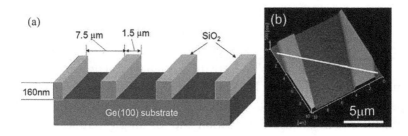

FIGURE 2.16 (a) Schematics of initial line-patterned SOG-SiO$_2$ /Ge structure. (b) AFM surface image. (Reproduced from S. K. Wang et al., *Journal of Applied Physics*. 108, 054104 (2010), with the permission of AIP Publishing.)

FIGURE 2.17 Cross-sectional profiles of as-received samples and those annealed at 600°C, oxygen partial pressure 0.1 Pa for 30, 90, and 150 min, respectively. (Reproduced from S. K. Wang et al., *Journal of Applied Physics*. 108, 054104 (2010), with the permission of AIP Publishing.)

Example 2.4: High-Pressure Oxidation of Ge

For many semiconductor materials, such as the Ge and III-V group of materials, the oxides formed by oxidation often have multiple states. The intermediate-state oxides are generally volatile. These volatile oxides are generally considered to be harmful and easily lead to the formation of defects, which in turn leads to the degradation of the quality and reliability of the dielectric. In this regard, it is necessary to consider suppressing the formation of these negative reactions.

According to the thermodynamic point of view, increasing the oxygen pressure helps to oxidize semiconductor to a higher valence state, thereby inhibiting the formation of those intermediate valence states. So high-pressure oxidation becomes a good way to avoid volatile reactions.

Here we take the oxidation of Ge as an example. Ge has two oxides, GeO and GeO$_2$. GeO is a relatively volatile one when compared with GeO$_2$, which is quite harmful to the formation of a high-quality gate dielectric. In order to suppress the GeO desorption during the thermal oxidation, high-pressure oxidation at >10 atm has been proposed and demonstrated to be effective in suppressing the formation of GeO.

Figure 2.18a and b show a schematic illustration and a camera illustration of the high-pressure oxidation system used for high-quality GeO_2 formation. It is mainly composed of a high-pressure oxygen cylinder, a pressure regulator, several valves, a rotary vacuum pump and a tube furnace, which consists of a quartz oxidation tube enclosed in a steel pressure vessel. The high-pressure oxidation system is evacuated to approximately 1 Pa by rotary pump after the cleaned Ge wafers are placed into the quartz oxidation tube. After that, the furnace chamber surrounding the steel pressure vessel is

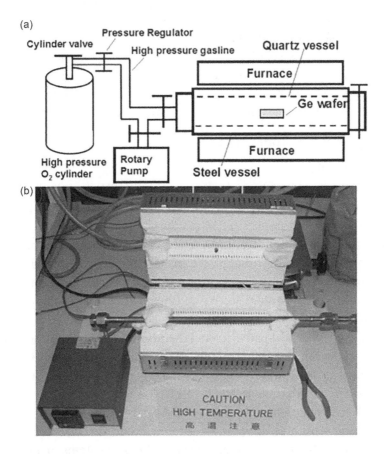

FIGURE 2.18   (a) Schematic illustration of HPO system; (b) HPO system used in this study.

heated to a thermal oxidation temperature. Temperature calibration of the high-pressure oxidation furnace was carried out in the temperature range from 200°C to 600°C for precise measurements.

2.2.2 Summary of Oxidation

Oxidation as the most basic insulating layer manufacturing method has been playing a huge role in modern semiconductor technology. In the new historical period, oxidation blooms more gorgeous flowers by introducing new oxidation species and a larger pressure range. For oxidant species, highly activated species are used instead of molecular oxygen, such as oxygen plasma, ozone, oxygen radical, to gain a low temperature, better thickness control or increase the dielectric quality. Moreover, as the novel requirements increase for advanced MOS devices, the oxidation pressure range has been extended a lot. Sometimes it is required to suppress the reaction that forms volatile materials for a high-quality gate dielectric, whereas sometimes active oxidation is used for surface modification or size control. No matter what we need, controlling the oxygen pressure from a thermodynamic point of view will always be the key.

2.3 DEPOSITION PROCESS

2.3.1 Sputtering

Sputter deposition is a physical vapor deposition (PVD) method of depositing thin films by sputtering a block of source material (target) onto a substrate. Sputtered atoms ejected into the gas phase are not in their thermodynamic equilibrium state and tend to deposit on all surfaces in the vacuum chamber. A substrate placed in the chamber will be coated with a thin film as shown in Figure 2.19. Sputtering sources are usually magnetrons that utilize strong electric and magnetic fields to trap electrons close to the surface of the magnetron, which is known as the target. The electrons follow helical paths around the magnetic field lines undergoing more ionizing collisions with gaseous neutrals near the target surface than would otherwise occur. The sputter gas is inert, typically argon. Under the action of an external electric field, electrons accelerate to fly out, collide with Ar atoms, ionize Ar atoms into Ar ions and secondary electrons, and transfer most of their energy to Ar ions. After Ar ions obtain energy, they bombard the target at high speed, making the atoms or molecules on the target fall off the surface of the target and splash out. These atoms or molecules that obtain energy fall on the surface of the substrate and

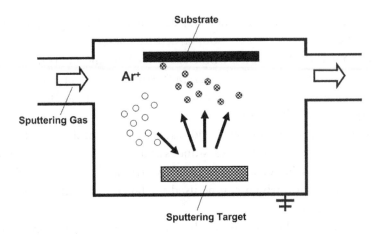

FIGURE 2.19   Illustration of the principle of sputtering.

precipitate to form particles coating. However, due to the occurrence of energy transfer many times, the electrons cannot bombard the ionization target, but directly fall on the substrate.

The extra argon ions created as a result of these collisions lead to a higher deposition rate. It also means that the plasma can be sustained at a lower pressure. The sputtered atoms are neutrally charged and so are unaffected by the magnetic trap. When sputtering gate dielectrics, insulating targets are used. Charge build-up on insulating targets can be avoided with the use of RF sputtering, where the sign of the anode–cathode bias is varied at a high rate. RF sputtering works well to produce highly insulating oxide films but only with the added expense of RF power supplies and impedance matching networks.

One important advantage of sputtering as a deposition technique is that the deposited films have the same composition as the source material. The equality of the film and target stoichiometry might be surprising since the sputter yield depends on the atomic weight of the atoms in the target. One might therefore expect one component of an alloy or mixture to sputter faster than the other components, leading to an enrichment of that component in the deposit. Sputter deposition also has an advantage over molecular beam epitaxy (MBE) due to its speed. The higher rate of deposition results in lower impurity incorporation because fewer impurities are able to reach the surface of the substrate in the same amount of time. Sputtering methods are consequently able to use process gases with far higher impurity concentrations than the vacuum pressure that MBE methods can tolerate. During sputter deposition, the substrate may be bombarded by energetic ions and neutral atoms. Ions can be deflected with a substrate bias and neutral bombardment can be minimized by off-axis sputtering, but only at a cost in deposition rate. Plastic substrates cannot tolerate the bombardment and are usually coated via evaporation.

Using this method, not only oxides such as Al_2O_3 and SiO_2 can be prepared, but also carbides, nitrides, sulfides and composite oxides. Therefore, sputtering is considered a convenient, powerful and safe way for depositing gate dielectrics, which can be used for fabricating various binary or ternary insulators.

By controlling the pressure of the reactive gas in the process of reactive sputtering, the films with different components can be obtained. The deposited products can be an alloy solid solution, compound or a mixture of them. However, with the increase of active gas pressure, a layer of corresponding compounds may be formed on the surface of the target during sputtering, resulting in the decrease of sputtering and film deposition rate. Reactive sputtering can be either DC reactive sputtering or RF reactive sputtering. Co-sputtering is to use two cathode targets made of different materials to sputter at the same time. By adjusting the sputtering discharge current on the cathode target, the composition of the film can be changed. It can also be achieved by fixing or pasting other material sheets on the surface of the main target.

Example 2.5: Sputtering of GeO_2-Solve the Oxygen Deficiency Problem

Here is an example that uses the RF sputtering method for fabricating GeO_2 films. And the argon gas flow rate is 22 sccm. The sputter power is changed from 20 to 100 W in order to tune the deposition rate. Figure 2.20 shows the sputtering system we used in this study. For GeO_2 film deposition, using only argon gas usually results in an oxygen-deficient GeO_2 because the deposition rates for Ge and O atoms are different. To overcome this problem, adding additional O_2 flow during the deposition process will be helpful. In this case, we add 0.6 sccm O_2 flow together with 22 sccm Ar flow during the GeO_2 deposition.

FIGURE 2.20   Typical sputtering system (ULVAC).

Similar to the above example, for most oxides, oxygen deficiency is a common problem. This will lead to the existence of a large number of oxygen vacancies, which will lead to the deterioration of the quality of the oxide layer and the increase of dielectric leakage. Therefore, adding additional O_2 or N_2 flow during the deposition process is an effective way to put oxygen or nitrogen into the dielectric network to suppress the gate leakage problem.

Example 2.6: Form Back Ohmic Contact to Si by Sputtering

Sputtering from the substrate rather than the target is also possible by reverse electrical connection. This is usually used to remove the native oxides or other contaminations on the semiconductor surface. For Si, the sputtering etching of the silicon layer shows damage extending to the wafer from 40 to 110 Å. The silicon surface may contain up to 20% of the atomic Ar, depending on the bias conditions. Material removal from the wafer surface can also be nonuniform, resulting in the formation of steep cones and etched valleys. However, for a typical sputtering etching, the required etching depth is less than 100 Å. For each sputtering system, the development of sputtering pre-cleaning steps must be experimentally optimized by measuring the impact of sputtering cleaning steps on contact resistance, contact reliability and junction leakage.

Some other useful tips:

What if the sputtering doesn't produce a glow? Sometimes it is the problem of the target surface; try opening the chamber and blow the target surface with an electric blower with ion wind.

How to deal with anoxia of sputtered materials? Add some oxygen; you can adjust the flow ratio to try. It should be noted that too much oxygen flow may affect the stability of ionization.

How to deal with insufficient target slots? Try the target sticking technology. Try to embed a small target into a large target to form a composite target.

What if some materials are difficult to glow? For the multi-target system, try to excite ionization at the target which is easy to produce a glow, and then open the shutter to guide the glow to the target which is difficult to produce a glow.

About pre-sputtering of target: It is usually required to clean the target before the deposition begins. For pre-sputtering, the plasma is ignited before opening the shutter, so that the material at the top of the target is deposited on the back of the shutter, rather than on the substrate.

2.3.2 Atomic Layer Deposition

Atomic layer deposition (ALD) is a kind of thin-film preparation technology suitable for the development of the latest and cutting-edge products. ALD is also an effective method for nanotechnology research. The typical application of ALD is to deposit high-precision, pinhole-free and shape-preserving thin films on substrates of various sizes and shapes. ALD is a chemical vapor deposition (CVD) technology. It was initially used to produce a nanostructured insulator (Al_2O_3/TiO_2) and zinc sulfide (ZnS) light-emitting layer of thin-film electroluminescent display (TFEL). Thanks to the development of ALD technology, this kind of display began mass production in the mid-1980s. The unique properties of ALD technology and the high repeatability of the process are the key factors in promoting the success of industrial production.

ALD is based on surface control. In the coating process, two or more chemical vapor precursors react on the substrate surface in turn to produce a solid film. In most ALD systems, a cross-flow reaction chamber is used, in which an inert carrier gas passes through and the precursor is injected into the inert carrier gas by a very short pulse. The inert carrier gas carries the precursor pulse as an orderly "wave" through the reaction chamber, vacuum pump pipeline, filtration system and finally through the vacuum pump. For the ALD process, the pressure range is generally about 0.1–10 mbar or atmospheric pressure with the temperature range of 50°C–500°C.

The most commonly used materials available for ALD include

Oxides: Al_2O_3, Ga_2O_3, HfO_2, La_2O_3, MgO, Nb_2O_5, Sc_2O_3, SiO_2, Ta_2O_5, TiO_2, V_xO_y, Y_2O_3, Yb_2O_3, ZnO, ZrO_2, etc.;

Nitrides: AlN, GaN, TaN_x, $TiAlN$, TiN_x, etc.;

Carbide: TaC, TiC, etc.; Metal: Ir, Pd, Pt, Ru, etc.;

Sulfide: ZnS, SrS, etc.;

Fluoride: CaF_2, LaF_3, MgF_2, SrF_2, etc.;

Biomaterials, polymers, and doped nano coating and composite structures.

Example 2.7: ALD in FinFET Process

With the emergence of 3D structures such as FinFET, omnidirectional ALD solutions are critical for dielectric, barrier and work function setting layers, and gate contacts. The maximum thermal budget is kept low, and theoretically, the metal deposition must be carried out below 500°C. The thermal ALD of pure metals is challenging in this temperature range, and most of the base metals that will form pure metals at this temperature are unstable and will mix impurities into the metal during deposition. However, the use of plasma-enhanced ALD (PEALD) has great advantages, so this technology can be used for low-temperature deposition of pure metals with the least amount of impurities. Both direct and remote plasmas can be used to deposit pure metals, but there are still some concerns about using plasmas near the gate. The industry continues to evaluate different low-temperature base metals to provide a solution for all temperature applications of pure metals deposited by ALD.

Example 2.8: Self-Cleaning Effect for III-V MOSFET

In 2001, Wilk and Ye et al. from Bell Lab first delivered ALD technology into the formation of high-k dielectrics on GaAs, opening a new era for III-V MOSFETs. By using alternating pulses of $Al(CH_3)_3$ and H_2O precursor in a carrier N_2 gas flow, the thickness and uniformity of the deposited oxide layer were well controlled at angstrom level. The growth of the ALD-Al_2O_3 process results in an abrupt interface with the GaAs substrate. The oxide layer appears as a desirable amorphous form, while the GaAs exhibits clear lattices. The ALD process removes the native oxide and excess As on the GaAs surface, resulting in a very thin Ga-Oxide interfacial layer. This process is now commonly acknowledged as the "self-cleaning" effect.

About Process and Film Properties

Excellent adhesion: The chemical adsorption between precursor and substrate ensures excellent adhesion.

Saturated adsorption characteristics: The self-limiting surface reaction makes it possible to automate the process without precise dose control and continuous operator intervention.

Ordered reaction: The digital ordered growth process of the thin film provides extremely high film precision without *in situ* feedback or operator intervention.

Surface-controlled reaction: Surface reaction ensures the high type retention of the film under any conditions, whether the substrate material is dense, porous, tubular, powder or other objects with complex shapes.

Accuracy and repeatability: The film growth thickness for one cycle is determined by the process, but it is usually 1 Å (0.1 nm).

Ultra-thin, dense and flat: ALD can deposit thin films with thickness less than 1 nm. In some industrial applications, the film thickness is only 0.8 nm.

High capacity: Surface-controlled growth makes it possible to expand capacity by increasing batch size and substrate area.

Plasma-enhanced ALD: Some metal, low-temperature oxide and nitride thin films can be prepared by adding plasma during ALD.

ALD for particles and powders: The combination of conformal coating and granular substrate creates many new applications, such as changing the diffusion characteristics of battery materials.

2.3.3 Vacuum Thermal Evaporation

Vacuum evaporation (including sublimation) is a PVD process where material from a thermal vaporization source reaches the substrate without collision with gas molecules in the space between the source and the substrate. The trajectory of the vaporized material is "line-of-sight". For an appreciable deposition rate to be attained, the material vaporized must reach a temperature where its vapor pressure is 10 mTorr or higher. Typical vaporization sources are resistively heated stranded wires, boats or crucibles (for vaporization temperatures below 1500°C). Figure 2.21a and b show the schematic and the camera photo of our vacuum evaporation system.

The metal layers of all early semiconductor technologies were deposited by evaporation. Although still widely used in some research applications and III-V technology, in most silicon technologies, evaporation has been replaced by other methods such as sputtering, ALD and so on. As indicated in Section 2.1 of the present chapter, for the preparation of small-scale MOS capacitors in the laboratory, the combination of the thermal evaporation method and the metal mask is a very effective and convenient method. It can simplify the preparation of electrodes as well as avoid the introduction of unnecessary contamination or damage.

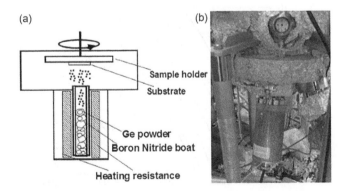

FIGURE 2.21   (a) Schematic and (b) camera photo of vacuum evaporation system.

The principle of the vacuum thermal evaporation method is that in a vacuum environment, the evaporation solid makes it into a gas state and is finally deposited on the surface of the sample. The vacuum here refers to the pressure in the container space of about 10^{-3} Pa. In this state, the gaseous particles of the evaporated solid raw materials can reach the surface of the sample smoothly. There are two most common heating methods in vacuum evaporation coating: resistance heating and electron beam heating. The resistance heating type is to heat the coating material by electrifying the resistance, which is generally applicable for metals with relatively low melting temperature, such as Al, Au, Ag. However, the electron beam heating type is to directly impact the electron beam emitted by the electron gun to the coating material, thus generating heat and achieving the purpose of heating the coating material. As the temperature of the sample rises, the material usually passes through solid, liquid and gas. At any temperature, there is an equilibrium vapor pressure above the material. When the sample falls below the melting temperature, this process is called sublimation. When the sample melts, it is called evaporation. Semiconductor process known as evaporation usually includes molten samples, but only because of the high vapor pressure in the operating area, which can produce an acceptable deposition rate. Figure 2.22 shows the equilibrium vapor pressure of the various elements as a function of temperature.

It is evident from Figure 2.22 that some materials must be heated to a much higher temperature than others to obtain the same vapor pressure. These materials, called refractory metals, include tantalum, tungsten,

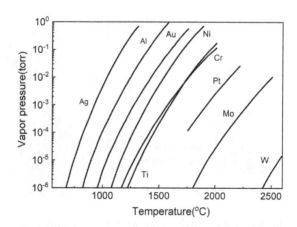

FIGURE 2.22   Vapor pressure curves for some commonly evaporated materials.

molybdenum and titanium, which have high melting points and, therefore, low vapor pressures at moderate temperatures. Tungsten requires a vapor pressure of more than 3000°C to reach 10 mTorr.

In order to obtain a large deposition rate, the evaporator usually operates at a very high crucible temperature. The vapor pressure above the crucible is high enough to keep the area in a viscous flow state. In this case, the operation will cause the vaporized material to condense into droplets. If these droplets move and attach to the surface of the wafer, the film will be nonuniform. For optimum uniformity, the evaporator must operate at low speeds, although a high vacuum is required to operate at low speeds to avoid fouling the film. The deposition rate is usually measured by a quartz crystal rate monitor. The device is a resonator plate that allows the resonant frequency to oscillate and then measures that frequency.

As mentioned earlier, one of the main limitations of evaporation is step coverage. In this case, the step is a contact section etched to the substrate by the insulating layer. At this distance scale (~ 1 μm), the incident beam can be considered as non-divergent. Assuming that the incident atoms are stationary on the surface of the wafer, this topology will produce definite shadows, while the film is usually discontinuous on the contact side. As the deposition proceeds, the growth layer at the top of the insulation layer moves up the shadow edge. For metallized layers, the coverage problem is particularly serious because these are the last steps in the process, and unless some planarization techniques are used, the accumulated topology can be very serious. Some limited improvements can be made by optimizing the position of the wafer. A common method to increase step coverage is to rotate the chips in the evaporation beam. Therefore, the hemispherical cage is placed on the wafer in the evaporator, which is designed to rotate the wafer on top of the evaporator. The deposition rate on the side wall is still lower than that on the plane, but it is uniform in the axial direction.

2.3.4 Molecular Beam Epitaxy (MBE)

MBE refers to the formation of a single crystal structure film on a single crystal substrate. The structure and orientation of the film are related to the structure and orientation of the substrate. If the substrate is the same as the film material, it is called homoepitaxy; otherwise, it is heteroepitaxy. MBE is similar to traditional vacuum evaporation technology in principle, but its equipment is huge and complex, including four main parts: UHV system, epitaxial growth system, *in situ* detection system and rapid sample exchange system. Because the epitaxial process is completed in an UHV, the scattering of the target molecular beam with residual gas and the superlattice structure produced by several Knudsen furnaces are avoided. MBE can control the orientation of molecules *in situ*, which can form two-dimensional ordered stacking on the atomic-level substrate. The vacuum of typical thin-film epitaxial growth technology ranges from 10^{-7} to 10^{-11} Torr. The epitaxial temperature is controlled by an electric furnace or Knudsen furnace, and the molecular beam formed by the evaporation source collimating through the small hole is aligned with the substrate perpendicular to it with a distance of 10–20 cm. The flow rate of the molecular beam is controlled by the furnace temperature. The furnace temperature is like a machine switch, which can control the flow rate of the molecular beam from on to off. By using this method of controlling the molecular beam flux continuously, we can also epitaxially grow multilayer quantum well-structured films with different compounds superimposed alternately. Typical epitaxial growth rates range from 0.001 to 100 Å/s. Because the thickness of the epitaxial ML is generally 3–5 Å, the corresponding growth rate is 0.7 ML/h to 30 ML/s. The lower limit of this range may make the rate of adsorbing impurities on the surface greater than that of adsorbing organic molecules, but it is difficult to control the epitaxial growth at too high a rate. For these reasons, the optimal growth rate should be controlled in the range of 0.1–5 Å/s. The monitoring of growth rate is similar to the principle of quartz crystal thickness monitoring in metal film deposition. During the film growth period, the temperature of the Knudsen furnace is often kept slightly lower than the sublimation temperature of the material in order to make the evaporation source continuously degassing. This method also needs a long time for a high-purity evaporation source. Therefore, in order to eliminate the impurities and water vapor in the evaporation source material stored in UHV, keeping it at a suitable temperature is the key factor, because these impurities and water vapor will lead to the formation of defects in the film growth.

One of the main parameters of epitaxial growth is the substrate temperature. There is a critical epitaxial temperature for a certain substrate and the corresponding film material. Above this temperature, the epitaxial growth is smooth and the film structure is good. Below this temperature, the epitaxial growth is imperfect. The typical substrate growth temperature range is 80–400 K, because the impurities will be adsorbed to the substrate surface quickly at low temperature. In the growth of organic films, the substrate temperature must ensure that the adhesion coefficient of molecules on the substrate surface is equal to 1, i.e., all molecules that strike the substrate surface must adhere to the intrinsic molecules on the surface. However, the substrate temperature used by some researchers is consistent with the temperature of adsorption rate, which can only control the ML growth of some organic molecular materials.

Figure 2.23 shows a diagram of a typical MBE system.

Recently, MBE has been used for the deposition of gate dielectrics using an *in situ* two-step method. Here is an example of the growth of wurtzite BeO on Ge (111).

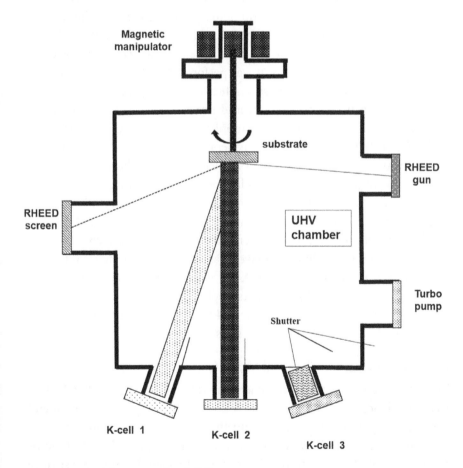

FIGURE 2.23   Schematic of an MBE system.

Example 2.9: Two-Step Growth of BeO on Ge (111)

Recently, by applying the ALD method, epitaxial BeO has been demonstrated to be a good choice for GaAs passivation because of its advantageous properties in the lattice (domain) match, large bandgap (10.6 eV), thermal stability and thermal conductivity. Since Ge has almost the same lattice constant with GaAs, BeO is presumably regarded as an ideal interface passivation layer on Ge. However, ALD-BeO precursor ($Be(CH_3)_2$) is highly toxic to harm researchers seriously, so in this example, we introduce a novel way for BeO epitaxially grown on Ge (111) with less security risk that may open a wide prospect for Ge surface passivation.

P-Ge (111) substrate was used in this example; after an HF-last cleaning process, the wafer was immediately transferred into the load-lock of RF-MBE system. Then, the substrate was thermally cleaned in a high vacuum at 650°C for 30 s to remove the native oxide followed by the deposition of Be layer. It was performed on Ge (111)-1×1 surface at 200°C, followed by *in situ* radical oxidation process.

Reflection high-energy electron diffraction (RHEED) technique was efficiently utilized to *in situ* monitor the whole growth process. On Ge (111)-1×1 clean surface, as depicted in Figure 2.24, a thin Be layer was first deposited onto the surface. It is found that the Be layer has a 30° in-plane rotation of its lattice with respect to the substrate by a 3:1 domain matching. In this way, the in-plane lattice misfit can be lowered to ~0.8%, which benefits the formation of high-quality Be film. Oxidation with active oxygen radicals was then performed, and a thin single-crystalline BeO layer was formed. To further confirm the structure of the BeO film, X-ray diffraction measurement was performed under θ–2θ mode. As depicted in Figure 2.25, according to another reference, it is inferred that the film is dominated by wurtzite BeO (002) under tensile stress.

2.3.5 Metal Organic Chemical Vapor Deposition (MOCVD)

The principle of metal organic chemical vapor deposition (MOCVD) is that metal organic CVD reaction source material (precursor of metal organic compounds) transforms into the gas state at a certain temperature and enters the CVD reactor with the carrier gas (H_2, Ar). One or more source materials entering the reactor diffuse to the substrate surface

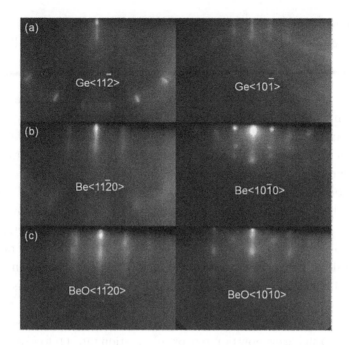

FIGURE 2.24 RHEED pattern of (a) as-cleaned Ge (111) surface; (b) after 650°C annealing, (1×1) reconstruction is obtained; (c) after metallic beryllium deposition; (d) after O-radical oxidation that forms BeO.

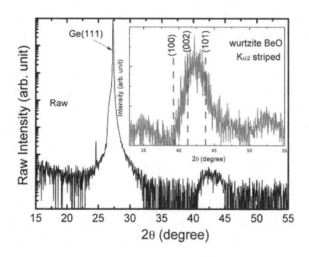

FIGURE 2.25 XRD spectrum of as-deposited BeO on Ge (111) substrate. The peak at around 42° is assigned to wurtzite BeO (002) with tensile stress.

through the vapor boundary layer, adsorb on the substrate surface and undergo one or more steps of the chemical reaction to epitaxially grow products; the gaseous reactants are discharged from the reaction system with the carrier gas.

MOCVD reaction is a nonequilibrium growth mechanism. The growth rate and microstructure of the epitaxial layer are affected by many factors, such as substrate temperature, reaction chamber pressure, metal organic precursor concentration, reaction time, substrate surface condition, gas flow property, etc.

The common precursors of CVD are metal hydrides, metal halides and metal organic compounds. Compared with metal hydrides and metal halides, metal organic compounds have a lower deposition temperature, a lower toxicity and corrosiveness to the reaction system, and most of them are volatile liquids or solids, which are easy to enter the reaction chamber with the carrier gas. The valuable organometallic compounds should have the following characteristics: (1) stable chemical properties at room temperature; (2) low evaporation temperature and high saturated vapor pressure; (3) stable evaporation rate or sublimation rate; (4) low decomposition temperature and suitable deposition rate (low deposition rate can be applied to the deposition of semiconductor films, and high deposition rate can be applied to the deposition of semiconductor films); (5) deposition of thick coating; (6) decomposition deposition process should not produce other impurities; (7) nontoxic, nonexplosive and spontaneous combustion, and the unreacted precursor should be easy to remove; (8) high purity and (9) low cost.

The simple MOCVD equipment shall have the following four basic systems: precursor supply system, reaction chamber, control system and tail gas treatment system. The reactor is the core part of the whole process of MOCVD. The mixture of reactants, the chemical reaction and the deposition of the products are all carried out in the reactor. Since the invention of the CVD method, the design of the reactor has been the focus and research object of CVD workers. The MOCVD reactor is mainly composed of a gas inlet device, a reaction chamber, a tray, a heater, an exhaust gas outlet and other parts. In the process of film growth by MOCVD, the gas is introduced into the reactor through the inlet device and flows through the tray, and a deposit reaction occurs on the substrate. Finally, it is discharged through the gas outlet to the reactor. According to the heating method of the reactor, the reactor of an MOCVD device can be divided into two types: hot wall type and cold wall type. The wall and substrate of the hot wall reactor are generally heated by a resistance heating furnace, which will cause the deposition of reaction products on the reactor wall. The cold wall heating methods include induction heating, electric heating and infrared heating. Cold wall heating can only be heated by the substrate itself, so deposition can only occur on the substrate, which is conducive to saving raw materials and processes. The MOCVD reactor can be divided into two types according to the flow direction of the main airflow relative to the substrate: a horizontal reactor with the main airflow parallel to the substrate direction and a vertical reactor with the main airflow perpendicular to the substrate direction.

Recently, MOCVD has been used as an *in situ* surface passivation method, followed by the growth of semiconductor epitaxy layers. Here is an example of the growth of SiN_x on GaN for surface passivation.

Example 2.10: MOCVD SiN_x Deposition on III-Nitrides

In this section, we take the deposition of silicon nitride amorphous film by MOCVD as an example to introduce how to deposit SiN_x film on GaN by MOCVD. SiN_x is often used as a passivation layer material in the semiconductor manufacturing process to achieve the purpose of isolation, protection or adjusting the surface morphology of the device, so as to further improve the performance of the device. NH_3 is the N source, and SiH_4 is the Si source.

The thickness of *in situ* SiN_x was measured by cross-sectional transmission electron microscopy (TEM). Then the thickness is normalized with the corresponding growth time to obtain the theoretical deposition rate. Figure 2.26a–c show the relationship between deposition rate as functions of growth temperature, pressure and SiH_4/NH_3 ratio. When the temperature increases from a lower temperature, the deposition rate increases sharply due to the increase in the reaction rate. However, further increase of temperature enhances the parasitic vapor reaction and the premature decomposition of SiN_x, resulting in the decrease of deposition rate at a higher temperature. This temperature dependence is different from that of SiN_x low-pressure chemical vapor deposition (LPCVD) using the SiH_4-NH_3 system, because the mass transport is limited at a higher deposition temperature of MOCVD. The relationship between deposition rate

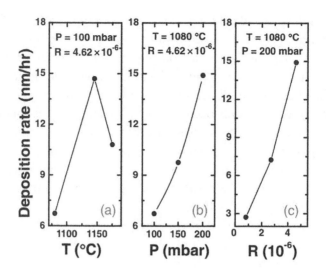

FIGURE 2.26 The relationship between deposition rate and (a) growth temperature, (b) growth pressure, (c) SiH_4/NH_3 ratio.

and pressure and SiH_4/NH_3 ratio is more direct. This is consistent with the reported LPCVD phase of SiN_x.

In addition to the growth rate, it is also crucial to understand how the material quality varies with the main growth parameters. SiH_4, the reaction source of SiN_x, will corrode group III nitrides, resulting in a poor uniformity of passivation layer and a rough interface and surface. When SiN_x first begins to deposit, discontinuous porous films are formed. In the hole positions of these undeveloped SiN_x, SiH_4 will corrode the lower class III nitrides, and then the porous film acts as the mask for SiH_4-selective etching of group III nitrides. The corrosion pits formed will be covered by the SiN_x coating deposited later. During the passivation process, group III nitrides under the passivation layer will also decompose at the holes, thus forming uneven interfaces and surfaces.

By promoting the lateral growth of SiN_x, a better interface can be obtained, so as to obtain a better coverage uniformity in the early stage. The decrease of pressure can improve the mobility of adsorbed atoms and promote the lateral growth of SiN_x. Higher NH_3 partial pressure can further promote the lateral growth of SiN_x, inhibit the decomposition of group III nitrides and reduce the SiH_4/NH_3 ratio, which can also alleviate the etching effect of SiH_4.

2.4 SUMMARY

Dielectric deposition process plays a key role in the preparation process of MOS devices, which is very important to improve the electrical properties of the MOS structure. This section covers several commonly used process technologies and explains the matters needing attention and application scope of these technologies through specific examples.

Overall, according to our experience, if there is a need to fabricate oxides with complex component control, especially for the fundamental research in the laboratory, it is recommended to use RF-sputtering as an initial method because it is not limited by the material species of the precursor and is more convenient to maintain.

If there is a need to fabricate dielectrics with a large size and uniform thickness or on a nonplanar surface, ALD is recommended. However, it is necessary to point out that in the ALD method, due to the limit of the precursor, materials variety is not as wide as the RF-sputtering, and the process temperature compatibility should also be considered.

According to our experience, the thermal evaporation method combined with a metal mask is the most convenient method for electrode preparation in the process of MOS capacitor manufacturing. Moreover, thermal evaporation is also an effective method to prepare oxide layers on the surfaces of materials lacking surface dangling bonds. For example, for carbon nanotubes or graphene, thermal evaporation of metal followed by *ex situ* oxidation has been widely used to form gate dielectrics on the surface.

In addition, we also introduce some methods of oxide epitaxy by MBE and surface passivation by MOCVD in this section, demonstrating the potential of these methods in the preparation of gate dielectrics. Due to the limited space and our experienced knowledge, some other important deposition methods, such as pulse-laser deposition (PLD), PECVD, low-pressure CVD (LPCVD), UHV-CVD, hydride vapor-phase epitaxy (HVPE), are not included in this book.

BIBLIOGRAPHY

1. Sze, S.M., and M.-K. Lee, *Semiconductor Devices: Physics and Technology*, 3rd Edition, (Wiley, New York, 2012).
2. Quirk, M., and J. Serda, *Semiconductor Manufacturing Technology*, (Prentice Hall, Englewood Cliffs, NJ, 2001).

3. You, N., X. Liu, J. Hao, Y. Bai, and S. Wang, Microwave plasma oxidation kinetics of SiC based on fast oxygen exchange, *Vacuum*, 2020. **182**: p. 109762.

4. Liu, X.-Y., J.-L. Hao, N.-N. You, Y. Bai, Y.-D. Tang, C.-Y. Yang, and S.-K. Wang, High-mobility SiC MOSFET with low DIT using high pressure microwave plasma oxidation, *Chinese Physics B*, 2020. **29**: p. 037301.

5. Liu, X., J. Hao, N. You, Y. Bai, and S. Wang, High-pressure microwave plasma oxidation of 4H-SiC with low interface trap density, *AIP Advances*, 2019. **9**(12): p. 125150.

6. Wang, S., and H. Liu, *Passivation and Characterization in High-k/III–V Interfaces, in Outlook and Challenges of Nano Devices, Sensors, and MEMS* (Springer, Cham, 2017), p. 123–149.

7. Yang, X., S.-K. Wang, X. Zhang, B. Sun, W. Zhao, H.-D. Chang, Z.-H. Zeng, and H. Liu, Al2O3/GeOx gate stack on germanium substrate fabricated by in situ cycling ozone oxidation method, *Applied Physics Letters*, 2014. **105**(9): p. 092101.

8. Wang, S.K., B.-Q. Xue, B. Sun, H.-L. Liang, Z.-X. Mei, W.Z. Hao, X.-L. Du, and H. Liu, *Extended Abstracts of the 2014 International Conference on Solid State Devices and Materials* 2012, p. 14–15.

9. Wang, S.K., B.-Q. Xue, B. Sun, H.-L. Liang, Z.-X. Mei, W.Z. Hao, X.-L. Du, and H. Liu, *IEEE International Conference on Solid-State and Integrated Circuit Technology* 2012.

10. Wang, S.K., K. Kita, C. H. Lee, T. Tabata, T. Nishimura, K. Nagashio, and A. Toriumi, Desorption kinetics of GeO from GeO2/Ge structure, *Journal of Applied Physics*, 2010. **108**(5): p. 054104.

CHAPTER 3

MOS Characterizations

3.1 METHODS FOR EVALUATING THE DENSITY OF INTERFACE STATES OF MOS

3.1.1 High-Frequency (Terman) Method

The high-frequency method is based on the high-frequency C–V curve to determine the interface trap density. It was proposed by Terman in 1962, so it is also called Terman method.

The content of the high-frequency method is when the frequency of small AC signal is high enough, the interface trap cannot respond to AC signal, so it has no effect on capacitance. However, the interface trap will change with the slow change of DC signal in the gate voltage. When the gate voltage changes, the interface trap will change, which will affect the bending degree of the energy band on one side of the semiconductor, making the C–V curve stretch along the voltage axis at high frequency, as shown in Figure 3.1.

For example, the n-type SiC metal–oxide–semiconductor (MOS) capacitor cannot respond to the small AC signal superimposed in the gate voltage at high frequency, and both of them have no contribution to the total capacitance of the MOS structure.

Therefore, the total capacitance of SiC MOS at high frequency is

$$C_{hf} = \left(\frac{1}{C_{ox}} + \frac{1}{C_s} \right)^{-1} = \left(\frac{1}{C_{ox}} + \frac{1}{C_a + C_d} \right)^{-1} \qquad (3.1)$$

DOI: 10.1201/9781003216285-4

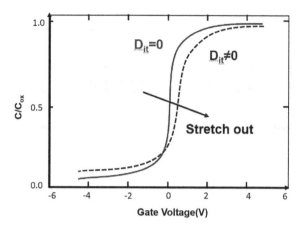

FIGURE 3.1 High-frequency $C–V$ curves of MOS capacitors with and without interface traps.

where C_s represents the total capacitance of semiconductors and is the parallel value of hole capacitance (C_a) and depletion layer capacitance (C_d). The corresponding equivalent circuit diagram is shown in Figure 3.2.

Then the total charge at one end of the gate electrode is

$$Q_G = -(Q_a + Q_d + Q_{it}) \tag{3.2}$$

The corresponding gate voltage can be expressed as:

$$V_G = V_{FB} + \varphi_s + \frac{Q_G}{C_{ox}} = V_{FB} + \varphi_s - \frac{Q_a + Q_d + Q_{it}}{C_{ox}} \tag{3.3}$$

If both sides of the equation are derived from φ_s at the same time, then we can get the equation as:

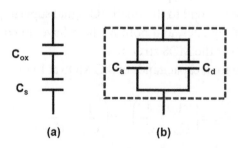

FIGURE 3.2 (a) SiC MOS equivalent circuit at high frequency and (b) equivalent circuit of C_s.

$$\frac{dV_G}{d\varphi_s} = 1 + \frac{1}{C_{\mathrm{ox}}}(C_a + C_d + C_{\mathrm{it}})$$ (3.4)

Further, we can get

$$C_{\mathrm{it}} = C_{\mathrm{ox}}\left(\frac{dV_G}{d\varphi_s} - 1\right) - C_s$$ (3.5)

According to the relationship between the density of states and the trap capacitance,

$$D_{\mathrm{it}} = \frac{C_{\mathrm{it}}}{q^2}$$ (3.6)

The interface density of states can be obtained from the high-frequency C–V curve

$$D_{\mathrm{it}} = \frac{C_{\mathrm{ox}}}{q^2}\left(\frac{dV_G}{d\varphi_s} - 1\right) - \frac{C_s}{q^2}$$ (3.7)

If we know the relationship between the gate voltage, V_G, and the surface potential, φ_s, we can calculate the change trend of the density of states with the surface potential according to Equation (3.7). According to the assumption of the high-frequency method, the interface trap can only respond to the DC signal but not to the AC signal at high frequency. Therefore, for a given surface potential, the ideal high-frequency capacitance is the same as the experimental one. By comparing the ideal high-frequency C–V curve with the experimental high-frequency curve, the corresponding relationship between the gate voltage and the surface potential can be obtained.

3.1.2 Quasi-Static (Low-Frequency) Method

The quasi-static C–V method is a commonly used method to measure the density of states (D_{it}). However, this method can only calculate the density of interface defects, and cannot provide the capture cross section and time constant of defects.

The basic theory of the quasi-static method is proposed by Berglund. This method mainly uses the quasi-static (low-frequency) C–V curve and the ideal C–V curve without interface state. When the test frequency is high enough, the ideal C–V curve can be approximated by the high-frequency C–V curve. "Low frequency" refers to the test frequency that the interface trap can respond to, so in the depletion region, the low-frequency capacitance (C_{lf}) can be expressed as:

$$C_{\text{lf}} = \left(\frac{1}{C_{\text{ox}}} + \frac{1}{C_s + C_{\text{it}}} \right)^{-1} \tag{3.8}$$

Then, we can get the expression of D_{it} from Equation (3.6)

$$D_{\text{it}} = \frac{1}{q^2} \left(\frac{C_{\text{ox}} C_{\text{lf}}}{C_{\text{ox}} - C_{\text{lf}}} - C_s \right) \tag{3.9}$$

When C_s is regarded as an ideal semiconductor capacitor, Equation (3.9) is applicable to the calculation of D_{it} in the whole bandgap range. However, the C_{lf} obtained in the test is the capacitance varying with V_G, and C_s is the capacitance varying with the surface potential (φ_s), so it is necessary to establish the relationship between V_G and φ_s. Berglund pointed out that the relationship between the two is

$$\varphi_s = \int_{V_{G1}}^{V_G} \left(1 - \frac{C_{\text{lf}}}{C_{\text{ox}}} \right) dV_G + \Delta \tag{3.10}$$

where Δ is the integral constant, which is related to the position of V_{G1}, when $V_{G1} = V_{\text{FB}}$, $\Delta = 0$; therefore, the relationship between V_G and φ_s can be obtained by determining V_{FB}.

The method to determine V_{FB}: V_{FB} can be obtained by the C–V curve; when the surface potential is zero, MOS capacitor is flat-band capacitor, and the flat-band voltage can be further calculated

$$C_{\text{FB}} = \frac{1}{\dfrac{t_{\text{ox}}}{\varepsilon_0 + \varepsilon_{\text{ox}}} + \dfrac{L_D}{\varepsilon_0 \varepsilon_s}} = \frac{C_{\text{ox}}}{1 + \dfrac{\varepsilon_{\text{ox}} L_D}{\varepsilon_s t_{\text{ox}}}} \tag{3.11}$$

$$L_D = \sqrt{\frac{\varepsilon_0 \varepsilon_s k_B T}{q^2 N_A}} \tag{3.12}$$

According to Equations (3.11) and (3.12), we can get

$$C_{FB} = \frac{C_{ox}}{1 + C_{ox} \Big/ \sqrt{N_A q^2 \varepsilon_0 \varepsilon_s / k_B T}} \qquad (3.13)$$

where C_{ox} is the dielectric capacitance, N_A is the substrate doping concentration, L_D is the Debye length, q is the electron charge, ε_0 is the permittivity of free space, t_{ox} is the dielectric thickness, ε_{ox} is the oxide dielectric constant, ε_s is the semiconductor dielectric constant, and k_B is Boltzmann's constant.

3.1.3 High–Low-Frequency Method

It is time-consuming to calculate the ideal C–V and the relationship between V_G and φ_s when calculating D_{it} from Equation (3.9). Therefore, Castagné and Vapaille proposed to use the high- and low-frequency method to measure D_{it}. The high–low-frequency method does not need to calculate the ideal semiconductor capacitance C_s, but uses the high-frequency capacitance (C_{hf}). This is because when the test frequency is high enough, the interface defects cannot respond. At this time, the measured capacitance value is the ideal capacitance value. Therefore, without considering the series resistance, the high-frequency capacitance is the semiconductor capacitance after removing the C_{ox} in series

$$C_s = \frac{C_{ox} C_{hf}}{C_{ox} - C_{hf}} \qquad (3.14)$$

The relationship between D_{it} and high–low-frequency capacitance can be obtained by introducing Equation (3.14) into Equation (3.9)

$$D_{it} = \frac{1}{q^2} \left(\frac{C_{ox} C_{lf}}{C_{ox} - C_{lf}} - \frac{C_{ox} C_{hf}}{C_{ox} - C_{hf}} \right) \qquad (3.15)$$

The change of D_{it} with the gate voltage can be calculated from Equation (3.15).

3.1.4 C–φ_s Method

There are two main problems in the application of high–low-frequency methods in SiC: (1) there are still interface traps to respond to 1 MHz AC signal, and the requirement of ohmic contact is too high for 100 MHz frequency; (2) when calculating the surface potential, the integration is usually started from V_{FB}, which is usually obtained by the high-frequency curve. If the high-frequency curve is not close enough to the ideal $C-V$ curve, the calculated surface potential will have some errors. Yoshioka et al. improved the high–low-frequency method and obtained the $C-\varphi_s$ method. In this method, the calculated ideal semiconductor capacitance is used to replace the semiconductor capacitance calculated by the high-frequency $C-V$ curve, and the interface density of states is obtained by comparing it with the semiconductor capacitance, including interface trap capacitor C_{it} under quasi-static condition.

In the $C-\varphi_s$ method, the relationship between the surface potential and the gate voltage is obtained by calculating the indefinite integral instead of the definite integral from V_{FB}

$$\varphi_s = \int \left(1 - \frac{C_{QS}}{C_{ox}} \right) dV_G + A \tag{3.16}$$

where A is the integral constant, which is included in the corresponding relationship between the surface potential and the gate voltage.

In this method, if the AC signal frequency of high-frequency $C-V$ is high enough, the interface state will not contribute to the MOS capacitance when the semiconductor is in the depletion state. The capacitance C_{dep} of the depletion region can be approximated by $(C_D + C_{it})$, where $C_{it} = 0$. Since the capacitance of the depletion layer is determined by the surface potential of the semiconductor, the relationship can be expressed as:

$$\frac{1}{(C_D + C_{it})^2} \approx \frac{1}{C_{dep}^2} = -\frac{2\varphi_s}{S^2 \varepsilon_{SiC} e N_D} \tag{3.17}$$

where S is the gate electrode area, ε_{SiC} is the relative dielectric constant of SiC material, N_D is the doping concentration of SiC epitaxial layer, C_D is the depletion layer capacitance, and C_{it} is the interface state capacitance. It can be seen from Equation (3.17) that the reciprocal of the square of the depletion layer capacitance has a linear relationship with the surface potential of the semiconductor, that is, the curve of $1/(C_D + C_{it})2 - \varphi_s$ should go through the origin.

Then, we can get the value of the integral constant A, first, we draw the $1/(C_D + C_{it})^2 - \varphi_s$ curve, and the corresponding relationship between

φ_s and V_G can be obtained by integrating the C–V curve. The obtained surface potential φ_s is plotted with the semiconductor capacitance of the high-frequency C–V curve. The straight line should pass through the origin. The value of constant A can be obtained through the intercept of the straight line on the horizontal axis, so as to obtain the relationship between the surface potential and the gate voltage. In addition, the value of N_D can be obtained by the slope of the line. Figure 3.3 shows the $1/\left(C_D + C_{it}\right)^2 - \varphi_s$ curve at different frequencies.

It can be seen from Figure 3.3 that for the same SiC MOS capacitor, when the AC signal frequency is 1 and 100 MHz, the gap of the depletion capacitance C_{dep} is very small, almost negligible.

Therefore, in the test process, it is enough to use the depletion layer capacitance of the 1 MHz high-frequency curve, which greatly reduces the requirements of test equipment and device technology. Under the condition that the relationship between φ_s and V_G is known, the depletion region semiconductor capacitance C_D can be calculated under ideal conditions:

$$C_{D,\text{theory}}(\varphi_s) = \frac{SeN_D \left|\exp\left(\dfrac{e\varphi_s}{KT}\right) - 1\right|}{\sqrt{\dfrac{2KTN_D}{\varepsilon_{SiC}}\left\{\exp\left(\dfrac{e\varphi_s}{KT}\right) - \dfrac{e\varphi_s}{KT} - 1\right\}}} \tag{3.18}$$

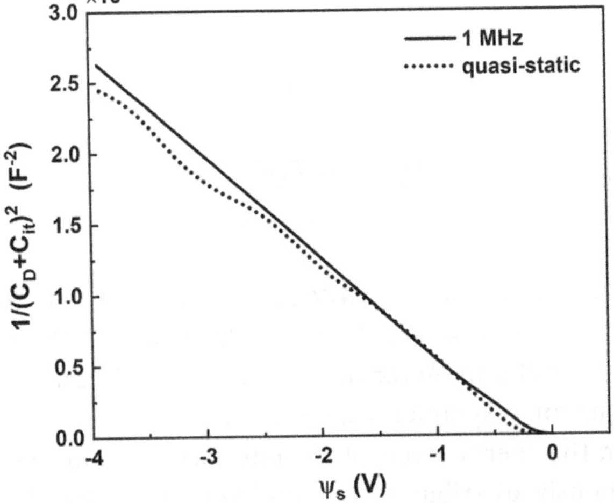

FIGURE 3.3   For n-type SiC MOS, $1/(C_D+C_{it})^2$–φ_s curves at different frequencies are obtained.

Comparing the calculated $C_{D,\text{theory}}$ with the experimental $(C_D + C_{it})$ under quasi-static conditions, the density of states at the interface is

$$D_{it} = \frac{(C_D + C_{it})_{QS} - C_{D,\text{theory}}}{Se^2} \tag{3.19}$$

3.1.5 Conductance Method

The conductance method was proposed by Nicollian and Goetzberger in 1967. The conductivity method is considered to be one of the most accurate methods to measure D_{it}. This technology is based on measuring the change of parallel conductance G_p with the gate voltage and frequency. The detectable D_{it} is as low as $10^9 \text{cm}^{-2}/\text{ev}$. The conductance method is also the most complex D_{it} test method at present, which can calculate the capture cross section, time constant, and surface potential fluctuation.

The conductance method mainly measures the parallel capacitance C_p and the conductance G_p of MOS capacitors. The simplified equivalent circuit is shown in Figure 3.4. It includes oxide capacitor, C_{ox}, semiconductor capacitor, C_s, and interface trap capacitance, C_{it}. The process of carrier capture/release by interface trap is an energy loss process, which is expressed by resistance, R_{it}. In order to facilitate the analysis and calculation, Figure 3.4a is usually simplified to the form of capacitor conductance parallel connection, namely, Figure 3.4b, where C_p and G_p are, respectively,

$$C_p = C_s + \frac{C_{it}}{1 + (\omega\tau_{it})^2} \tag{3.20}$$

$$\frac{G_p}{\omega} = \frac{q\omega\tau_{it}D_{it}}{1 + (\omega\tau_{it})^2} \tag{3.21}$$

where $\omega = 2\pi f$, f is the test frequency, τ_{it} is the time constant corresponding to the interface trap, $\tau_{it} = R_{it}C_{it}$. Equations (3.20) and (3.21) are only for the case of single-level trap. According to Equation (3.21), it can be seen that the G_p/ω equation is symmetric about $\omega\tau_{it}$.

However, the energy levels of the interface traps in the SiC bandgap are continuously distributed. For the continuous interface traps, the electron capture and release mainly occur in the range of several $k_B T/q$ near the Fermi level. In this case, the time constant of the trap is no

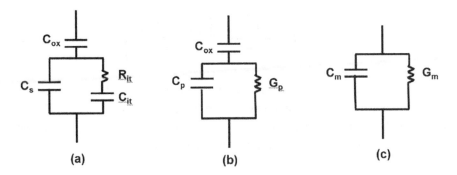

FIGURE 3.4 Equivalent circuit tested by the conductance method: (a) MOS capacitance of the interface trap time constant ($\tau_{it}=R_{it}C_{it}$); (b) simplified circuit diagram of (a); and (c) actual measurement circuit.

longer a fixed value, but will be discretized. At this time, the normalized conductance G_p/ω is

$$\frac{G_p}{\omega} = \frac{qD_{it}}{2\omega\tau_{it}}\ln\left(1+(\omega\tau_{it})^2\right) \qquad (3.22)$$

It can be seen from Equation (3.22) that when ω is $2/\tau_{it}$, there is a maximum value of G_p/ω, and the corresponding D_{it} value is $2.5G_p/q\omega$. However, Equation (3.22) does not consider the influence of the surface potential fluctuation on the G_p/ω-f curve in SiC. Considering the large surface potential fluctuation in SiC, the G_p/ω-f curve obtained in the experiment is wider than that obtained by Equation (3.22). There are many factors that cause the surface potential floating, such as the uneven distribution of the charge at the oxide layer and the interface, the uneven doping concentration and the thickness of the oxide layer near the interface, etc., the randomly distributed charge at the interface will change the electric field strength near its position, and then cause the disturbance of the surface potential.

Considering the effect of the surface potential fluctuation, the expression of conductance is

$$\frac{G_p}{\omega} = \frac{q}{2}\int_{-\infty}^{\infty}\frac{D_{it}}{\omega\tau_{it}}\ln\left[1+(\omega\tau_{it})^2\right]P(U_s)dU_s \qquad (3.23)$$

where $P(U_s)$ represents the probability distribution of surface potential perturbation.

$$P(U_s) = \frac{1}{\sqrt{2\pi\sigma_s^2}} \exp\left(-\frac{\left(U_s - \bar{U}_s\right)^2}{2\sigma_s^2}\right)$$
(3.24)

where $\bar{U}_s$ is the normalized mean surface potential and σ_s is the standard deviation of the probability distribution. Comparing the circuits in Figures 3.4a and c, the value of G_p/ω can be obtained from the capacitance C_m and conductance G_m obtained from the actual test.

$$\frac{G_p}{\omega} = \frac{\omega G_m C_{ox}^2}{G_m^2 + \omega^2 \left(C_{ox} - C_m\right)^2}$$
(3.25)

In the practical application of the conductance method, the influence of the parasitic series resistance of test systems and test equipment on test data cannot be ignored. In practice, the circuit diagram considering parasitic resistance is shown in Figure 3.5. When MOS is in the strong accumulation region at high frequency, there is no depletion layer on one side of the semiconductor. At the same time, the interface traps are short-circuited by most carrier capacitors. At this time, the equivalent circuit is only the series connection of oxide trap and parasitic resistance R_s. Furthermore, it is deduced that R_s is

$$R_S = \frac{G_{ma}}{G_{ma}^2 + \omega^2 C_{ma}^2}$$
(3.26)

C_{ma} and G_{ma} represent the capacitance and conductance values measured in the strong accumulation region, respectively. After R_s is extracted,

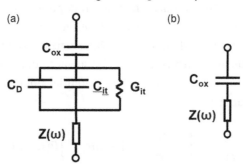

FIGURE 3.5 The equivalent circuit diagram of MOS C–V tests under different states including parasitic resistance: (a) from depletion to the accumulation region and (b) the strong accumulation region.

the measured capacitance and conductance can be corrected. The corrected capacitance and conductance are represented by G_c and C_c, respectively,

$$G_c = \frac{\left(G_m^2 + \omega^2 C_m^2\right)a}{a^2 + \omega^2 C_m^2} \tag{3.27}$$

$$C_c = \frac{\left(G_m^2 + \omega^2 C_m^2\right)C_m}{a^2 + \omega^2 C_m^2} \tag{3.28}$$

$$a = G_m - (G_m^2 + \omega^2 C_m^2 R_S) \tag{3.29}$$

The corrected C_c and G_c are brought back to Equation (3.25), and the value of D_{it} can be obtained by using the new G_p/ω.

3.2 EXPERIMENTAL STEP

Experimental apparatus: KEYSIGHT E4990A impedance analyzer (Figure 3.6).

LCR measurement principle

The meter does not directly measure the values of L, C and R. Instead, it calculates the impedance or admittance of the DUT by measuring the change of the measured signal (a known signal passes through the measured part, and the output signal will change), and then calculates the corresponding L, C and R values according to the impedance or

FIGURE 3.6   KEYSIGHT E4990A impedance analyzer.

admittance by selecting different measurement modes. There are two measurement modes of LCR, series mode and parallel mode. The series mode is based on measuring impedance Z ($Z=U/I$), and the parallel mode is based on measuring admittance Y ($Y=I/U$). The schematic diagram is shown in Figure 3.7.

In Figure 3.7, V_s is the internal sine wave standard signal source of the instrument, and R_s is the internal resistance of the instrument. V is the internal digital voltmeter of the instrument, and A is the internal digital ammeter of the instrument.

After the measured part is connected to the measuring instrument, the digital voltmeter will measure the voltage at both ends of the measured part, the digital ammeter will measure the current flowing through the measured part, and the phase detector will measure the angle between the voltage and the current. When the measurement mode is selected, the instrument can calculate the parameters to be measured by internal computer according to the values of U, I and θ. The parameters that the instrument can detect are shown in Figure 3.8.

The meanings of these parameters are as follows: Z is the impedance, Y is the admittance, R is the resistance, G is the conductance, C_p is the parallel capacitor, C_s is the series capacitor, L_p is the parallel inductance, L_s is the phase angle of voltage relative to the current, X is the reactance, B is the admittance, D is the loss factor, Q is the quality factor, R_p is the parallel resistance, and R_s is the series resistance.

3.2.1 Calibrate the Equipment

Calibration reason: calibration eliminates the parasitism associated with the system wiring and probe, and reduces the influence of error sources between the wafer device and the instrument, "calibration plane"—it defines the calibration plane of the wafer probe tip. For low-frequency calibration (< 1 MHz), open-/short-circuit calibration is sufficient.

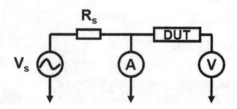

FIGURE 3.7   Schematic diagram of LCR principle.

parameters

FIGURE 3.8   Parameters that can be measured by E4990A.

4TP: four BNC to SSMC cables, L_{pot}, L_{cur}, H_{pot} and H_{cur}, are required for LCR/Z instrument connection. The connection of 4TP 1 m BNC to SSMC provides better high-frequency performance above 80 MHz. In some cases, a 2-m-long cable may be required to connect to the LCR/Z instrument. Calibration above 80 MHz may not be possible with a 2-m cable. Specific operations can be analyzed according to their own experimental conditions.

3.2.1.1 Phase Calibration

1. After opening E4990A, it needs at least 30 minutes to warm up; after preheating, click "preset", and then click "OK".

2. Then click "cal", select "adapter", and select "4TP 1m" or "4TP 2m" according to the actual situation.

3. Connect the calibration equipment E4990-61001 with the instrument; in other words, connect the L_{CUR} and L_{POT} terminals on the 16048G or 16048H to the H_{POT} and H_{CUR} terminals of the 100 Ω

resister (furnished with the E4990A, Part No. E4990-61001). Other terminals should remain open, as shown in Figure 3.9a.

4. Click Phase to start the phase compensation data measurement. When the phase compensation data measurement is completed, the softkey label changes to Phase [Done].

5. Click "Save Phase".

6. Connect the L_{CUR}, L_{POT}, H_{POT} and H_{CUR} terminals on the 16048G or 16048H to the L_{CUR}, L_{POT}, H_{POT} and H_{CUR} terminals of the 100 Ω resister as shown in Figure 3.9b.

7. Click E4990-61001 to start the measurement.

8. When the load data measurement is completed, the softkey label changes to E4990-61001 [Done].

9. Click Save Impedance to start calculating the adapter setup data from the measured phase compensation and load data. The adapter setup data are automatically saved to the E4990A (Figure 3.10).

After the phase calibration, connect the H_{pot} and H_{cur} ports of the instrument output end with a T-type converter and convert them into one output. The same operation is used to convert the L_{pot} and L_{cur} outputs into one output, as shown in Figure 3.11.

BNC joint is divided into two types: one is called coaxial joint, and the other is called triaxial joint; the actual structure is shown in Figure 3.12; the structure diagram is shown in Figure 3.13.

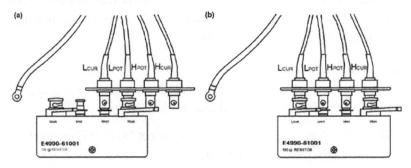

FIGURE 3.9 Connection modes between calibration equipment e4990-61001 and instrument. (a) Phase compensation data measurement. (b) Load compensation data measurement.

(a)

(b)

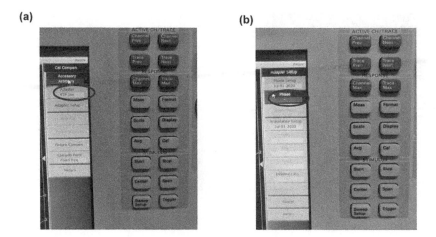

FIGURE 3.10   Steps of phase calibration. (a) Click adapter setup. (b) Save phase compensation data to the E4990A.

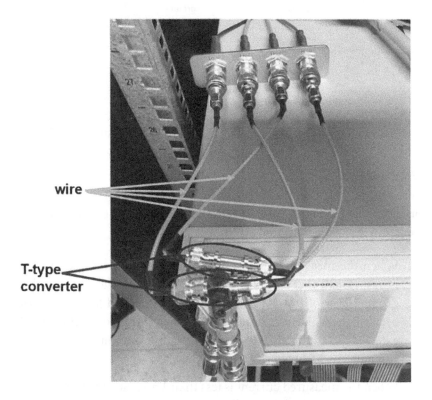

FIGURE 3.11   Conversion of output port by T-type converter; the wire length is generally about 25 cm.

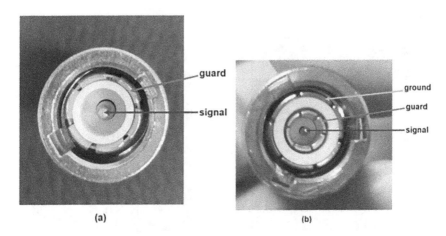

FIGURE 3.12 The actual structure of BNC joint: (a) coaxial joint and (b) triaxial joint.

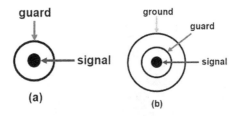

FIGURE 3.13 The structure diagram of BNC joint: (a) coaxial joint and (b) triaxial joint.

3.2.1.2 Butt Joint of Coaxial Joint and Triaxial Joint

There will be an insulating layer in the coaxial and triaxial conversion ports, as shown in Figure 3.14. Its function is to separate the coaxial guard layer from the triaxial ground layer.

1. When the measured current is greater than 1 nA, the guard layer is suspended, that is, the signal is connected to the signal and the ground layer of triaxial is connected to the guard layer of coaxial, because the leakage current between the central conductor and the external ground shield will not have a significant impact on the measurement, as is shown in Figure 3.15a.

2. When the measured current is less than 1 nA, the only way to maintain the accuracy of the small current measurement is to connect the guard layer of triaxial with the guard layer of coaxial, which makes

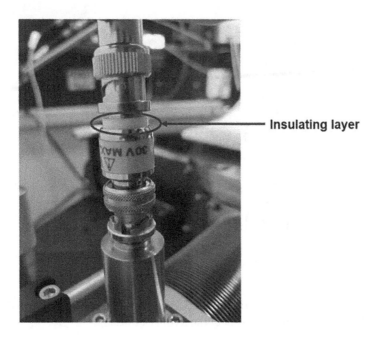

Insulating layer

FIGURE 3.14   Insulating layer in the coaxial and triaxial conversion port.

the central signal surrounded by the guard layer, and the BNC outer shielding layer cannot be grounded and must be floating. The voltage of the BNC shielding layer can be 100 or 200 V. This brings potential safety hazards, so it is necessary to be very careful in the measurement settings to ensure that no accidental electric shock will occur. Special connectors and adapters are usually required to implement this connection, as is shown in Figure 3.15b.

In the process of wiring, it is necessary to connect the guard layers of two BNC connectors with wires, which are called feedback lines. The outer

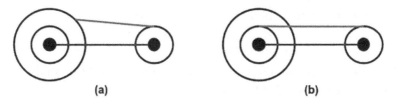

(a)   (b)

FIGURE 3.15   Butt joint of coaxial joint and triaxial joint: (a) signal–signal, ground layer of triaxial-guard layer of coaxial and (b) signal–signal, guard layer of triaxial-guard layer of coaxial.

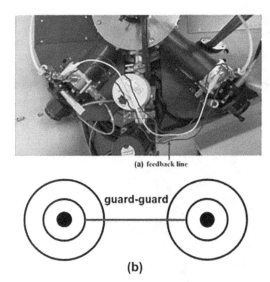

FIGURE 3.16 The display of the feedback line: (a) real product of the feedback line and (b) the structure diagram of the feedback line.

shield of the BNC cables coming from a capacitance meter is actually not at ground potential, but is "virtual grounds". In order to stabilize the inductance of the cables, it is important to supply a "return path" through the BNC shield. The wiring is shown in Figure 3.16, the white line is the feedback line in Figure 3.16a, and the structure diagram of the feedback line is shown in Figure 3.16b.

After the wire is connected, the sample to be measured is gently placed on the sample holder of the probe table with tweezers, as is shown in Figure 3.17. In the actual measurement process, in order to eliminate the signal-to-noise ratio, we adopt the reverse side method, that is, the gate is connected to the low level, and the substrate is connected to the high level, as shown in the figure. After the probe table is sealed, open-circuit calibration and short-circuit calibration are carried out.

3.2.1.3 Open-Circuit Calibration
In the case of open circuit, the resistance is infinite, and the conductivity is zero, so the measured value is C_p–G.

1. Hang the two selected probes in the air (do not contact the sample holder or sample), that is, it is equivalent to open circuit.

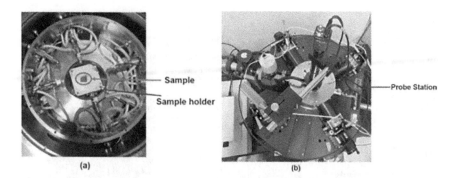

FIGURE 3.17 (a) Sample holder and (b) probe station.

2. Then press "cal" key, select "fixture compen", and click "open" for open-circuit calibration.

3. Next click "meas" key to select the measured value of trace as C_p–G, and the corresponding curve will be displayed.

3.2.1.4 Short-Circuit Calibration
In case of short circuit, the resistance is zero, so the measured value is R–X.

1. Make the two selected probes to be both in contact with the sample holder, that is, in the short-circuit state, the distance between the probes can be adjusted closer.

2. Then click "cal", select "fixture compen", and click "short" to calibrate the short circuit.

3. Next click "meas" to select the measured value of trace as R-X, and the corresponding curve will be displayed.

Steps of open calibration and short calibration are shown in Figure 3.18.

3.2.2 C–V Curve Was Measured After Calibration
1. Set the value of the small AC signal: click "sweep setup", and then click "OSC level" to set the AC voltage value to 50 mV, then click "return" to return to the sweep menu.

2. Set scan type: click "sweep type", if the DC voltage is scanned, then select "DC bias"; if the frequency is scanned, then select "log

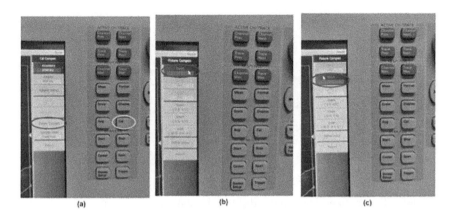

FIGURE 3.18   Steps of open calibration and short calibration: (a) click "fixture compen", (b) click "open", and (c) click "short".

frequency". Then click "span" on the instrument to set the scanning range. Here, we take scanning voltage as an example.

3. Set frequency: select "CW freq" in the sweep setup menu to set the frequency of the small AC signal.

4. Set bandwidth: click the "avg" button and select "meas time" to select the bandwidth, which is generally selected as 3. It depends on the experiment.

   The bandwidth is the scanning accuracy. The value ranges from 1 to 5, which represents the scanning times of each point. The larger the number, the more the scanning times, the higher the scan is, the more accurate the data will be, but the longer it will take.

5. Set trigger type: click "trigger" button, and then select "trigger source". If it is controlled by an external computer, click "external"; if it is measured directly by the instrument, select "internal".

6. Click the "display" on the instrument to adjust the number of coordinate axes.

After all the above are set, the physical quantity can be measured.

   One of the two selected probes is in contact with the sample holder and the other with the MOS to be measured. According to the above steps, the C_p–G curve can be adjusted, and the interface state density of the corresponding MOS can be calculated according to the conductance method.

3.2.3 An Example of Measuring Density of Interface States of SiC MOS by Conductance Method

First, the instrument is calibrated according to the steps in Section 3.2.1, and then, interface traps measurement is started.

3.2.3.1 Part 1: Measurement of the C–V Curve

The number of interface states in response to different frequencies is different; the C–V curve will change accordingly. The change caused by this frequency is called dispersion. It means that there is an interface state when there is dispersion in the C–V curve. Therefore, the voltage range of dispersion can be determined by measuring the C–V curve under varying frequencies.

The method of measuring the C–V curve at different frequencies is as follows:

1. Set the value of the small AC signal: click "sweep setup", select "OSC level", and set the value of the small AC signal voltage to 50 mV, after setting, click "return" to back to the sweep menu.

2. Then click "CW freq" to set the frequency of the small AC signal, the frequency range of this experiment is 100 Hz–1 MHz, and the frequency setting values of each measurement are 100, 1000 Hz, 10,000, 100,000 Hz and 1 MHz, after setting the frequency, click "return" to back to the sweep menu.

3. Next select "DC bias", press the "span" button on the instrument to input the DC bias range, this experiment is −5–8 V, click "return" to back to the sweep menu.

4. Set bandwidth: click "avg" button, select "meas time" and select 3.

5. Set trigger type: click "trigger" button, select "trigger source", and select internal trigger.

6. One of the two selected probes is in contact with the sample holder and the other with the MOS to be measured.

7. Click "meas" and select the "C_p–G" axes.

Repeat the above steps and change the frequency only for each measurement. The V_G range corresponding to dispersion can be known by comparing the quintic C–V curve, as shown in Figure 3.19. By calculating the

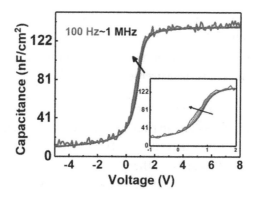

FIGURE 3.19 The C–V curve of 100 Hz–1 MHz, the small picture is the enlarged figure of the dispersion area. (Reproduced from X. Y. Liu et al., *AIP Advances* 9, 125150 (2019), with the permission of AIP Publishing.)

flat-band voltage, the V_G range corresponding to the MOS in the depletion region and the dispersion appears can be obtained.

3.2.3.2 Part 2: Measurement of the G–f Curve

A new sample is used to measure the conductance, and a certain step size is set in the known V_G range. In this case, we scan in the V_G range of 1.72–1.88 V with a step size of 0.02 V; the C_m_logf and G_m_logf data under each fixed V_G can be obtained.

The method to measure the conductance under fixed V_G is as follows:

1. Set the value of the small AC signal: click "sweep setup", then click "OSC level" to set the value of the small AC signal voltage to 50 mV; after setting, click "return" to back to the sweep menu.

2. Click "sweep type", select the scanning type of the frequency as "log freq".

3. Click the "span" button on the instrument to set the initial value and final value of the scanning frequency, which are 100 Hz and 1 MHz, respectively, or set the scanning frequency curve by computer program.

4. Set the value of V_G, click "sweep setup", select "DC bias", then click "voltage level" to input fixed V_G value; the value of each V_G represents a surface potential; each experiment can measure the corresponding information of V_G.

5. Set bandwidth: click "avg" and select meas time, select 3;

6. Set trigger type: click "trigger" button and select "trigger source", because it is controlled by computer program; "external trigger" is selected in this experiment.

7. Click "meas" and select the "C_p–G" axes.

The G–f curve under the fixed surface potential can be obtained by only changing the value of V_G in each measurement. Since the temperature will affect the time constant of the interface trap, the measurement is carried out at $T = 100\,\text{k}$.

3.2.3.3 Part 3: Measurement of the System Series Resistance R_s

R_S is measured in the high-frequency strong accumulation region, so the fixed gate voltage V_G of SiC MOS in the strong accumulation region in 1 MHz C–V data is used as the fixed gate voltage. In a fixed V_G, the impedance $Z = R + jX$ of the system at different frequencies is measured. The obtained R and X values are linearly fitted in the high-frequency range (10 kHz–1 MHz). The intercept of the fitting curve on the R-axis is the series resistance R_s of the system, as shown in Figure 3.20.

After getting the value of R_S, we bring it into Equation (3.29) to get the value of a, where G_m and C_m in Equation (3.29) are the data directly

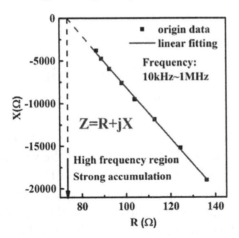

FIGURE 3.20 Derivation process of parasitic resistance R_s of test system. (Reproduced from Z. Y. Peng et al., *Journal of Applied Physics* 123, 135302 (2018), with the permission of AIP Publishing.)

obtained from the instrument. Then, we bring the parameter a into Equations (3.27) and (3.28) to get the modified G_m and C_m values, namely, G_C and C_C. If we bring the values of G_C and C_C into Equation (3.25), we can get the value of G_p/ω without the influence of R_s.

Data Analysis

Calculation of the oxide voltage C_{ox}: for SiC MOS, the semiconductor capacitance C_s of SiC MOS at high frequency is much larger than that of oxide capacitor C_{ox}. Therefore, the maximum value C_{max} of the C–V curve at 1 MHz is C_{ox}.

Calculation of the surface potential φ_s corresponding to V_G: the maximum frequency of this experiment is 1 MHz, so the high-frequency C–V curve is the C–V curve at 1 MHz. According to this curve, we can get the minimum capacitance $C_{min, HF}$; at this time, the depletion layer reaches the maximum width (W_{dm}), and the expression of semiconductor capacitance is

$$\frac{1}{C_{min,HF}} = \frac{1}{C_{ox}} + \frac{1}{C_S} \tag{3.30}$$

where C_{ox} is known, and C_s can be expressed as Equation (3.31):

$$\frac{1}{C_S} = \frac{W_{dm}}{\varepsilon_s \varepsilon_0} \tag{3.31}$$

$$W_{dm} = \sqrt{\frac{4\varepsilon_s \varepsilon_0 k_B T \ln(N_A / n_i)}{N_A q^2}} \tag{3.32}$$

where ε_s is the dielectric constant of the semiconductor—in this case, it is the dielectric constant of SiC, $\varepsilon_s = 9.7$; ε_0 is the vacuum dielectric constant; k_B is the Boltzmann constant; and N_A is the substrate doping concentration. Substituting Equation (3.32) into (3.31) can obtain N_A. Next, the flat-band voltage can be calculated from the flat-band capacitor, and the expression of flat-band capacitance is shown in Equation (3.33):

$$\frac{1}{C_{FB}} = \frac{1}{C_{ox}} + \frac{1}{C_S} \tag{3.33}$$

C_{ox} is known, and C_s can be expressed as Equation (3.34):

$$\frac{1}{C_S} = \frac{L_D}{\varepsilon_s \varepsilon_0} \tag{3.34}$$

where L_D is the Debye length, and its expression is

$$L_D = \sqrt{\frac{\varepsilon_s \varepsilon_0 k_B T}{N_A q^2}} \tag{3.35}$$

The Debye length, L_D, can be obtained by introducing N_A into Equation (3.35), and the flat-band capacitor, C_{FB}, can be obtained. Therefore, the corresponding flat-band voltage, V_{FB}, can be found on the 1 MHz C–V curve; then the surface potential is calculated by Equation (3.10), when $V_{G1} = V_{FB}$, $\Delta = 0$; select the C–V data with lower frequency, take 100 Hz as an example, export the C–V data with $f = 100$ Hz, that is to say, C_{lf} is obtained, and transform it into $(1 - C_{lf}/C_{ox})$, and then V_G is integrated as follows:

$$\varphi_s = \int_{V_{FB}}^{V_G} \left(1 - \frac{C_{lf}}{C_{ox}} \right) dV_G \tag{3.36}$$

It should be noted that the gate voltage, V_G, should make the SiC MOS in the depletion region because the conductivity method can only measure the D_{it} of the depletion region with high accuracy. After obtaining φ_s, the corresponding energy range can be calculated according to the energy band diagram, that is, the value of $E_c - E_t$, where E_t is the energy level at a certain position in the forbidden band.

Then, the value of the interface states can be obtained by the following steps:

1. First, the range of σ_s is set, and the relationship between ξ_p and σ_s is obtained by formula (3.37), as shown in Figure 3.21.

$$\int_{-\infty}^{\infty} \exp\left(-\frac{\eta^2}{2\sigma_s^2} \right) \exp(-\eta) \left\{ \frac{2\xi_p^2 \exp 2\eta}{1 + \xi_p^2 \exp 2\eta} - \ln\left(1 + \xi_p^2 \exp 2\eta \right) \right\} d\eta = 0 \tag{3.37}$$

2. Then set the range of σ_s and take $n = 5$, according to the calculation results on the right side of Equation (3.38) or (3.39), Figure 3.22 can be obtained.

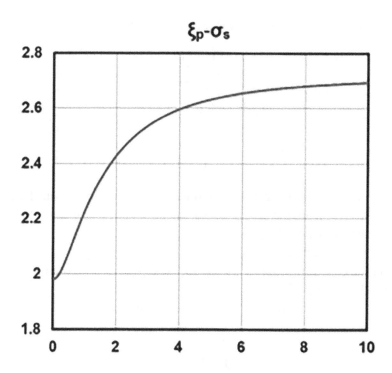

FIGURE 3.21 The relationship between ξ_p and σ_s.

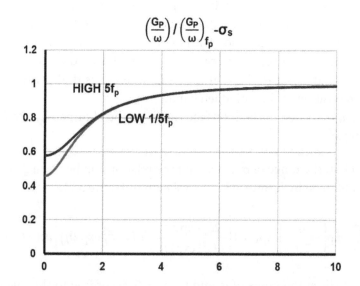

FIGURE 3.22 The diagram of the relationship between $(G_p/\omega)/(G_p/\omega)_{fp}$ and σ_s.

$$\frac{(<G_p>/\omega)_{f_p/n}}{(<G_p>/\omega)_{f_p}}=n\frac{\displaystyle\int_{-\infty}^{\infty}\exp\!\left(-\frac{\eta^2}{2\sigma_s^2}\right)\exp(-\eta)\ln[1+\xi_p^2\exp(2\eta)/n^2]d\eta}{\displaystyle\int_{-\infty}^{\infty}\exp\!\left(-\frac{\eta^2}{2\sigma_s^2}\right)\exp(-\eta)\ln[1+\xi_p^2\exp(2\eta)]d\eta} \qquad (3.38)$$

$$\frac{(<G_p>/\omega)_{nf_p}}{(<G_p>/\omega)_{f_p}}=\frac{1}{n}\frac{\displaystyle\int_{-\infty}^{\infty}\exp\!\left(-\frac{\eta^2}{2\sigma_s^2}\right)\exp(-\eta)\ln\!\left(1+n^2\xi_p^2\exp 2\eta\right)d\eta}{\displaystyle\int_{-\infty}^{\infty}\exp\!\left(-\frac{\eta^2}{2\sigma_s^2}\right)\exp(-\eta)\ln\!\left(1+\xi_p^2\exp 2\eta\right)d\eta} \qquad (3.39)$$

3. Next, the G_p/ω–$\log f$ curve can be displayed from Part 2; as shown in Figure 3.23, the frequency f_p corresponding to the peak value can be calculated according to the data in Figure 3.23.

4. For Equation (3.38) or (3.39), take $n=5$, the corresponding value on the left side of the equation can be found in Figure 3.23, that is, the ratio of G_p/ω corresponding to $5f_p$ or $1/5f_p$ to its peak value;

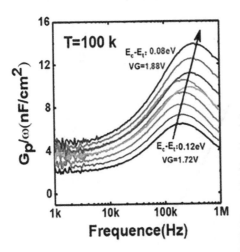

FIGURE 3.23 The diagram of G_p/ω–$\log f$, the value of V_G corresponding to blue line is 1.72 V, and the value of V_G corresponding to red line is 1.88 V. (Reproduced from X. Y. Liu et al., *AIP Advances* 9, 125150 (2019), with the permission of AIP Publishing.)

then, the standard deviation σ_s corresponding to each ratio can be obtained from Figure 3.22.

5. Then in Figure 3.21, the corresponding ξ_p can be obtained from the known value of σ_s.

6. Then according to Equation (3.40), the relation curve between f_D and σ_s can be obtained, as shown in Figure 3.24.

$$f_D(\sigma_s) = \frac{\left(2\pi\sigma_s^2\right)^{-1/2}}{2\xi_p} \int_{-\infty}^{\infty} \exp\left(-\frac{\eta^2}{2\sigma_s^2}\right) \exp(-\eta) \ln\left(1 + \xi_p^2 \exp 2\eta\right) d\eta \quad (3.40)$$

Find the value of f_D corresponding to the known σ_s in Figure 3.24, and then bring f_D into Equation (3.41) to obtain the corresponding value of D_{it}.

$$D_{it} = \left(\frac{G_p}{\omega}\right)_{fp} [f_D(\sigma_s)q]^{-1} \quad (3.41)$$

It can be seen from Figure 3.24 that the value of f_D is between 0 and 0.4, generally f_D is 0.4, so we can get the value of D_{it} is $D_{it} = 2.5 \frac{1}{q}\left(\frac{G_p}{\omega}\right)_{fp}$.

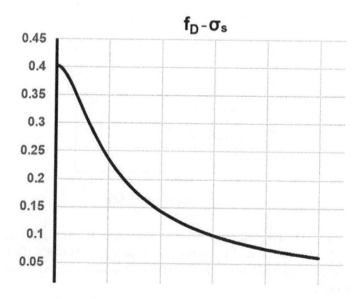

FIGURE 3.24 The diagram of the relationship between f_D and σ_s.

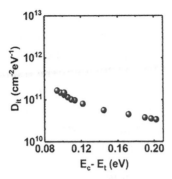

FIGURE 3.25 Relationship between density of interface states and energy levels of 4H SiC. (Reproduced from X. Y. Liu et al., *AIP Advances* 9, 125150 (2019), with the permission of AIP Publishing.)

Through the above analysis, the density of interface states at the fixed gate voltage, V_G, can be calculated, and the desired curve can be obtained by corresponding to the calculated energy-level difference $E_c - E_t$, as shown in Figure 3.25.

3.3 HYSTERESIS AND BULK CHARGE

At present, the biggest obstacle to the development of 4H-SiC MOSFET is that the mobility of the inversion layer is too low. The main reason for the low mobility is that there are a lot of interface state trapped charges, near interface oxide trapped charges and fixed charges in the oxide layer in the SiO_2/4H-SiC MOS system. These three kinds of charges are the causes of low threshold reliability, low mobility and transconductance difference of devices. Understanding the influence mechanism of these three kinds of charges on the device performance is helpful to improve the oxide interface quality process.

In fact, there are a lot of trapped charges in the oxide layer, including interface trapped charge, near interface trapped charge and oxide fixed charge. There are three main differences between the trapped charge at the interface and that in the oxide layer near the interface.

First, they are located differently in the device. The trapped charge is located at the interface of 4H-SiC/SiO_2, while the trapped charge of the oxide layer near the interface is in the oxide layer several nanometers away from the interface; second, the time constants are different, that is, the time required for them to exchange charges with the substrate is different. The trapped charge at the interface takes a short time and can follow the large DC signal of the gate voltage, so it is also called "fast state", while the trapped charge near the interface takes a long time to respond to the change of the gate voltage, which is also called "slow state"; third, their distribution in the bandgap is different. The trapped charge at the interface can be approximately regarded as a continuous distribution, which is generally considered as a "U"-type distribution, that is, the trap density near the edge of the conduction band or valence band increases, while that near the interface oxide layer is discontinuous.

3.3.1 Interface Trapped Charge

Because the time constant of the interface trap charge is small, it can follow the large signal of the gate voltage, which is also called "fast state". The interface trap charge will induce a trap capacitor, C_{it}, by trapping or releasing carriers, which can be regarded as the relationship between the interface trap charge and surface capacitance, C_s, in parallel. The equivalent circuit is shown in Figure 3.26b. At this time, the actual total capacitance can be expressed as follows:

$$\frac{1}{C} = \frac{1}{C_{ox}} + \frac{1}{C_s + C_{it}} \tag{3.42}$$

The gate voltage is

$$V_G = V_{FB} + \varphi_s - \frac{Q_s + Q_{it}}{C_{ox}} \tag{3.43}$$

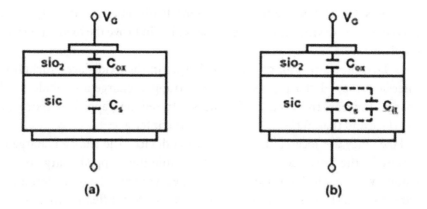

(a)   **(b)**

FIGURE 3.26 The MOS capacitor equivalent circuit (a) ideal case and (b) actual situation with trap.

The results show that the actual C–V curve is relatively ideal, and the C–V curve becomes smoother because the addition of the interface trap charge can capture and release electrons quickly, thus obtaining an interface trap capacitor, C_{it}, in series with the gate oxide capacitor, C_{ox}, so the total capacitance is smaller than the expected C_{ox}.

3.3.2 Near Interface Trapped Charge (Border Trap)

Due to the large time constant of the trapped charge near the interface, it is unable to respond to the high-frequency DC signal of the gate voltage. Therefore, the hysteresis phenomenon will appear in the C–V curve when the gate voltage is scanned forward and backward. In a certain range, the longer the gate voltage stress duration, the more obvious the hysteresis phenomenon. At present, the trap density of the near interface oxide layer is usually calculated by the hysteresis voltage, V_{hy}, as is shown in Equation (3.44):

$$N_{iot} = \frac{C_{ox} V_{hy}}{q} \tag{3.44}$$

Zhao-yang Peng et al. proposed a method to evaluate the near interface trap density (border trap) based on the high-frequency 1 MHz C–V curve.

Considering that the near interface traps in SiC devices seriously affect the reliability of devices, the equivalent circuit in Figure 3.27a is no longer applicable. They proposed the modified equivalent circuit in Figure 3.27b. In this circuit, we divide the oxide capacitance into several limited segments, and the contribution of the near interface trap is represented by a series of capacitance resistance circuits, and the corresponding time constant can be expressed as $\tau_{BT} = C_{BT} R_{BT}$. One end of the near interface trap is connected with the semiconductor, and the other end is connected at different depths of the oxide layer.

According to the modified equivalent circuit, we can conclude that the near interface trap will not only affect the measured capacitance, but also affect the surface potential of the semiconductor side. According to the equivalent circuit in Figure 3.27b, the parameters of near interface trap can be obtained by analyzing the C–V curve obtained from the test.

Before the C–V test, the parasitic resistance, R_s, of the test system is evaluated, Section 3.2.3.3 describes the detailed steps to obtain R_s.

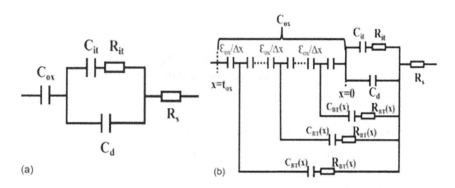

FIGURE 3.27 (a) The traditional equivalent circuit model during SiC MOS C–V test and (b) the modified equivalent circuit model considering the distribution of near interface traps in the oxide layer. C_{it}, R_{it}, C_{BT}, R_{BT} represent the additional capacitance and resistance caused by interface traps and near interface traps, respectively. (Reproduced from Z. Y. Peng et al., *Journal of Applied Physics* 123, 135302 (2018), with the permission of AIP Publishing.)

After eliminating the influence of parasitic resistance on the test data, the C–V test was started. The amplitude of the small AC signal used in the test was 50 mV.

It is worth noting that the near interface traps are mainly located in SiO_2 and have a larger time constant compared with the interface traps, so the 1 MHz C–V curve at room temperature will not contain the information of near interface traps. The time constant relationship between interface traps and near interface defects is shown in Figure 3.28. In order to detect the near interface trap by the 1 MHz AC signal, it is necessary to increase the response frequency of the near interface trap, that is to say, it needs to be tested at high temperature.

The premise of this method is that we assume that the time constant of the near interface trap follows the Shockley–Reed–Hall statistical model, that is to say,

$$\tau(T) \propto \exp\left(\frac{E_a}{k_B T}\right) \tag{3.45}$$

where E_a is the thermal activation energy of electron tunneling into the oxide layer, and k_B is the Boltzmann constant. It can be seen that with the increase in temperature, the time constant of the trap decreases, and the response frequency increases.

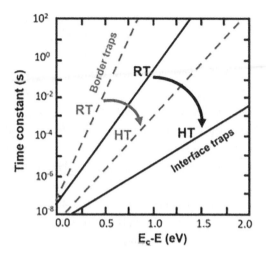

FIGURE 3.28 Comparison of time constants of interface traps and near interface defects at different temperatures. (Reproduced from Z. Y. Peng et al., *Journal of Applied Physics* 123, 135302 (2018), with the permission of AIP Publishing.)

Based on this, we test and obtain 1 MHz *C–V* curves at different temperatures, as shown in Figure 3.29a. It can be seen that with the increase in temperature, the dispersion between 1 MHz curves increases gradually, which is similar to the dispersion characteristics between different frequencies at

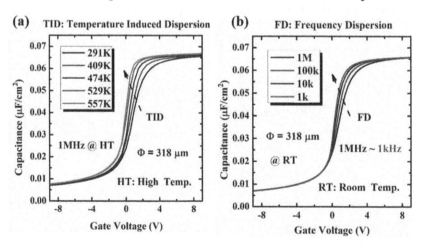

FIGURE 3.29 (a) 1 MHz *C–V* curves at different temperatures and (b) *C–V* curves at different frequencies at room temperature. (Reproduced from Z. Y. Peng et al., *Journal of Applied Physics* 123, 135302 (2018), with the permission of AIP Publishing.)

room temperature (Figure 3.29b). We define the dispersion between high temperature 1 MHz curves as temperature-induced dispersion. Comparing Figures 3.29a and b, it is easy to see that the 1 MHz curve at high temperature corresponds to the C-V curve at a low frequency at room temperature.

By comparing the dispersion at high temperature with that at room temperature (Figure 3.30), a 1 MHz curve corresponding to the 1 kHz curve at room temperature can be found. We define dispersion as the maximum capacitance difference between two C-V curves and normalize it to C_{ox}. The dispersion between 1 kHz and 1 MHz curves at room temperature is $0.2C_{ox}$. In the temperature-induced dispersion characteristics, it is found that the dispersion between 1 MHz at 438 k and 1 kHz at room temperature is $0.2C_{ox}$, which indicates that the 1 MHz curve at 438 k corresponds to the 1 kHz curve at room temperature.

In addition, it can be observed that the frequency dispersion between 557 and 438 k 1 MHz C-V curves is much greater than that between 1 kHz and QSCV curves, which indicates that the dispersion between 557 and 438 k has other sources. At this time, interface defects are no longer the main contributor of dispersion. We attribute this part of dispersion to the reaction of border traps, that is to say, when the temperature is higher than 438 k, the near interface trap is detected in the 1 MHz curve instead of the

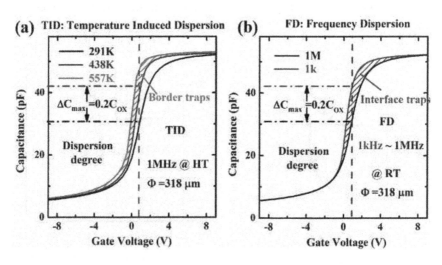

FIGURE 3.30 (a) Temperature-induced dispersion characteristics between 438 k, 557 k and 1 MHz curves at room temperature and (b) dispersion characteristics of 1 kHz and 1 MHz curves at room temperature. (Reproduced from Z. Y. Peng et al., *Journal of Applied Physics* 123, 135302 (2018), with the permission of AIP Publishing.)

interface state. The reason for this result is that the time constant of the trap near the interface increases with the increase in temperature, and the rate of electron release becomes faster, and the number of electrons trapped in the trap will decrease, which leads to the change of the hysteresis characteristics of the C-V curve.

Figure 3.31 shows the hysteresis characteristics of the 1 MHz C-V curve at high temperature and the corresponding hysteresis voltage versus temperature. With the increase in temperature, the hysteresis voltage of the curve increases gradually from the positive to the negative direction, which indicates that the hysteresis voltage increases in the negative direction, which proves that the near interface trap response increases gradually with the increase in temperature. The negative hysteresis voltage indicates that the near interface trap type is the acceptor type.

The near interface trap density can be obtained by the 1 MHz C-V curve with a temperature above 438 k. The near interface trap charge density between different depths can be obtained by integrating the difference of capacitance of two adjacent curves on the voltage axis at high temperature, as shown in Figure 3.32. The distribution diagram of the near interface trap charge density with its position in the oxide layer obtained by the above method is shown in Figure 3.33. The closer to the interface, the higher the trap density near the interface, the lower the trap density.

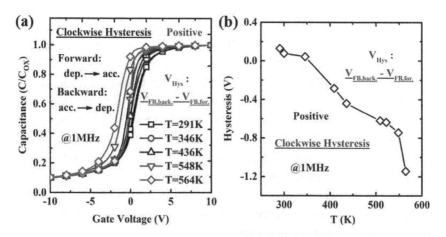

FIGURE 3.31 (a) Hysteresis characteristics of the 1 MHz C-V curve at high temperature and (b) temperature dependence of hysteresis voltage derived from hysteresis characteristics. (Reproduced from Z. Y. Peng et al., *Journal of Applied Physics* 123, 135302 (2018), with the permission of AIP Publishing.)

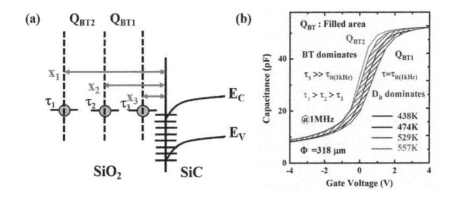

FIGURE 3.32 (a) The distribution of near interface traps in the oxide layer and (b) border trap charge Q_{BT} obtained. (Reproduced from Z. Y. Peng et al., *Journal of Applied Physics* 123, 135302 (2018), with the permission of AIP Publishing.)

3.3.3 Fixed Charge in the Oxide Layer

The fixed charge in the oxide layer will not capture or release electrons, so the *C–V* curve will not be delayed, which will only make the *C–V* curve drift to the left and right. For example, the positive fixed charge will induce the negative charge at the gate electrode, which will make the whole *C–V* curve drift to the negative gate voltage direction.

3.4 EQUIVALENT OXIDE THICKNESS

In order to realize the continued scaling of MOS devices, a suitable replacement for SiO_2 should be soon developed. After several years of intensive effort, a new material known as "high-κ" dielectric has been identified. High-κ stands for high dielectric constant, which is a measure of how much charge a material can hold. Different materials similarly have different abilities to hold charges. As an alternative to SiO_2, high-κ metal oxides can provide a substantially thicker (physical thickness) dielectric for reduced leakage and improved gate capacitance.

As for the gate capacitance issue, capacitance of a parallel plate capacitor can be expressed as (ignoring quantum mechanical and depletion effects from a Si substrate and gate):

$$C = \frac{\kappa \varepsilon_0 A}{t} \tag{3.46}$$

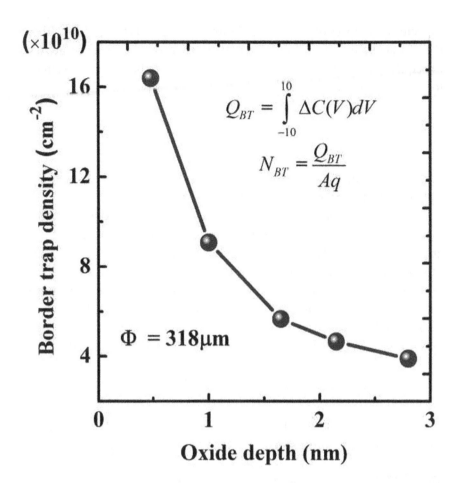

$$Q_{BT} = \int_{-10}^{10} \Delta C(V) dV$$

$$N_{BT} = \frac{Q_{BT}}{Aq}$$

$\Phi = 318\mu m$

FIGURE 3.33 Distribution of border traps with their depth in the oxide layer. The interface position of SiC/SiO$_2$ at $x=0$. (Reproduced from Z. Y. Peng et al., *Journal of Applied Physics* 123, 135302 (2018), with the permission of AIP Publishing.)

where κ is the dielectric constant (also referred to as the relative permittivity) of the material, ε_0 is the permittivity of free space (=8.85×10^{-3} fF/ μm), A is the area of the capacitor, and t is the thickness of the dielectric. This expression for C can be rewritten in terms of t_{eq} (i.e., equivalent oxide thickness (EOT)) and κ_{ox} (=3.9, the dielectric constant of SiO$_2$) of the capacitor. The term t_{eq} represents the theoretical thickness of SiO$_2$ that would be required to achieve the same capacitance density as the dielectric (ignoring issues such as leakage current and reliability). For example, if the capacitor dielectric is SiO$_2$, $t_{eq}=3.9\varepsilon_0(A/C)$, and a capacitance density of $C/A = 34.5$ fF/μm^2 corresponds to $t_{eq} = 1$ nm. Thus, the physical thickness of an alternative dielectric employed to achieve the equivalent capacitance density of $t_{eq} = 1$ nm can be obtained from the following expression:

$$\frac{t_{eq}}{\kappa_{ox}} = \frac{t_{high-\kappa}}{\kappa_{high-\kappa}} \qquad (3.47)$$

For example, a dielectric with a relative permittivity of 16, therefore, affords a physical thickness of ~4 nm to obtain $t_{eq} = 1$ nm.

Next, how to extract EOT from the experiment. Usually, EOT is extracted by capacitance, for example, making an on-silicon capacitance, growing a dielectric layer on a silicon substrate, and then growing a metal gate, and testing its capacitance–voltage curve, as shown in Figure 3.34. Then by fitting the experimental capacitance–voltage curve, the equivalent oxide layer thickness and flat-band voltage are obtained. There are commercial fitting software and free ones, such as Berkeley's QMCV program. It should be noted that when the equivalent oxide layer thickness

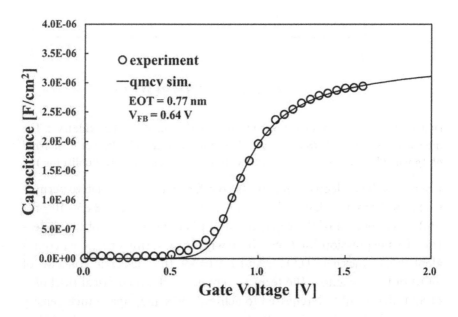

FIGURE 3.34 Experimental and theoretically simulated capacitance–voltage curves.

is less than 2 nm, the capacitor will not be saturated in the accumulation area, but will continue to increase as the gate voltage increases. This is due to the large capacitance of the gate stack, and the change of the substrate capacitance is more obvious.

3.5 LEAKAGE

The traditional MOS gate stack mainly consists of heavily doped polysilicon electrodes, SiO_2 dielectrics and silicon substrates. SiO_2 dielectric has served for more than three decades as the excellent gate insulator responsible for blocking the current in insulated-gate field-effect transistor channels from the gate electrode in CMOS devices. To improve the device performance, SiO_2 has been scaled aggressively to invert the surface to a sufficient sheet charge density to obtain the desired current for the given supply voltage and to avoid short-channel behavior. However, with the rapid scaling down of devices, SiO_2 is scaled from a thickness of 100 nm 30 years ago to a mere 1.2 at 90 nm process node. This represents an oxide layer composed of only four atoms thick. This tendency will cause a problem that as the SiO_2 layer gets thinner, the rate of gate leakage tunneling exponentially goes up. Current leakage contributes to power dissipation and heat. The relation of tunneling current leakage current with SiO_2 thickness can be obtained by the following equation:

$$J_g = \frac{A}{T_{ox}^2} e^{-2T_{ox}\sqrt{\frac{2m^*q}{\hbar^2}\left\{\Phi_B-\frac{V_{ox}}{2}\right\}}} \tag{3.48}$$

where A is an experimentally constant. T_{ox} is the physical thickness of the SiO_2 dielectric. ϕ_B is the potential barrier height between the metal and SiO_2. V_{ox} is the voltage drop across the dielectric. m^* is the electron effective mass in the dielectric. For highly defective films that have electron trap energy levels in the SiO_2 bandgap, electron transport will instead be governed by a trap-assisted mechanism such as Frenkel–Poole emission or hopping conduction. The dependence of the leakage current on SiO_2 physical thickness is shown in Figure 3.35. Consequently, this conduction problem causes the transistor to stray from its purely "on" and "off" state and into an "on" and "leaky off" behavior.

Figure 3.36 shows the gate leakage characteristics of a typical n-type silicon substrate gate structure. In the accumulation area of the silicon

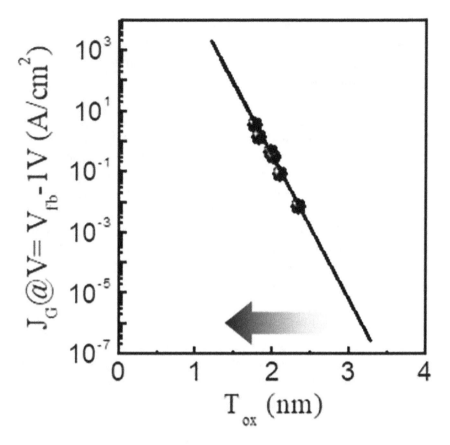

FIGURE 3.35 Gate leakage current density plot as a function of SiO_2 physical thickness.

substrate, the leakage current is large, which is caused by the reduction of the electron barrier.

3.5.1 Direct Tunneling

Next, let's discuss the leakage mechanism. First, we discuss the direct tunneling mechanism. In the direct tunneling mechanism, the tunneling current is related to the oxide layer thickness, T_{ox}, the effective electron mass, m_{ox}, of the oxide layer and the barrier height, ϕ_b. According to the experimental results, select the appropriate m_{ox} and ϕ_b to calculate all direct tunneling components: including electron tunneling in the conduction band, electron tunneling in the valence band and hole tunneling in the valence band. The direct tunneling model expression is

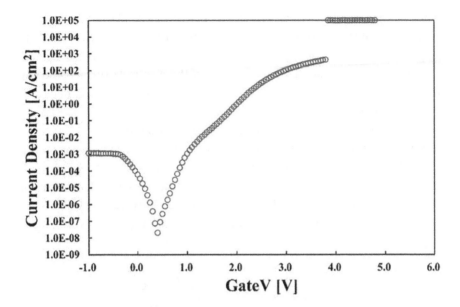

FIGURE 3.36 The gate leakage characteristics of a typical n-type silicon substrate gate structure.

$$J_{DT} = \frac{q^3}{8\pi h \phi_b \varepsilon_{ox}} \cdot C(V_g, V_{ox}, T_{ox}, \phi_b)$$

$$\exp\left(-\frac{8\pi\sqrt{2m_{ox}q}\phi_b^{\frac{3}{2}}\left[1-\left(1-\frac{V_{ox}}{\phi_b}\right)^{\frac{3}{2}}\right]}{3hE_{ox}}\right) \tag{3.49}$$

where q is the basic charge, h is the Planck constant, ϕ_b is the tunnel barrier height, ε_{ox} is the effective dielectric constant, T_{ox} is the physical thickness of the effective oxide layer, V_g is the gate voltage, V_{ox} is the voltage drop on the oxide layer, and M_{ox} is the equivalent effective mass of oxide electrons.

$$C(V_g, V_{ox}, T_{ox}, \phi_b) = \exp\left[\frac{20}{\phi_b}\left(\frac{|V_{ox}|-\phi_b}{\phi_{b0}}+1\right)^{\alpha}\cdot\left(1-\frac{|V_{ox}|}{\phi_b}\right)\right]\cdot\left(\frac{V_g}{T_{ox}}\right)\cdot N \tag{3.50}$$

where ϕ_{b0} is the effective barrier height. The exponential term in function C contains all the quadratic effects experimentally to fit the current characteristics of V_{ox} from $0\,V$ to ϕ_b. The secondary effect mainly considers the factors such as the density of states on the electrode surface and the unfixed effective mass in the oxide layer. The V_g term can ensure that when V_g is equal to $0\,V$, the current is zero.

The N in formula (3.50) has different meanings in different tunneling processes. For ECB (electron tunneling from conduction band) and HVB (hole tunneling from valence band) cases, it represents the number of carriers, and for EVB (electron tunneling from valence band) cases, it represents the tunneling probability. In general, N expresses the on-state of different tunneling processes. For ECB and HVB processes, N can be expressed as:

$$N = \frac{\varepsilon_{ox}}{T_{ox}} \left\{ n_{inv} v_t \cdot \ln\left[1 + \exp\left(\frac{V_{ge} - V_{th}}{n_{inv} v_t} \right) \right] \right.$$

$$\left. + n_{acc} v_t \cdot \ln\left[1 + \exp\left(-\frac{V_g - V_{FB}}{n_{acc} v_t} \right) \right] \right\} \qquad (3.51)$$

where V_{ge} represents the voltage after the depletion of polysilicon is removed, which can be obtained by analytical methods:

$$V_{ge} = V_{FB} + \phi_{S0} + \frac{q \varepsilon_{Si} N_{poly} T_{ox}^2}{\varepsilon_{ox}^2} \cdot \left(\sqrt{1 + \frac{2 \varepsilon_{ox}^2 (V_g - V_{FB} - \phi_{S0})}{q \varepsilon_{Si} N_{poly} T_{ox}^2}} - 1 \right) \qquad (3.52)$$

Here, the ECB process mainly occurs in the accumulation area and inversion area of NMOS and the accumulation area of PMOS. HVB mainly occurs in the inversion area of PMOS. For N in the EVB process, it can be expressed as

$$N = \frac{\varepsilon_{ox}}{T_{ox}} \left\{ n_{EVB} v_t \cdot \ln\left[1 + \exp\left(\frac{|V_{ox}| - \phi_g}{n_{EVB} v_t} \right) \right] \right\} \qquad (3.53)$$

When $V_{ox} < \phi_g$ the electrons of EVB tunnel to the other side, they are in the center of the forbidden band, which means that there is no density of states that accepts tunneling, and the tunneling process cannot occur. Therefore,

N here can be understood as the probability of tunneling. The EVB process can provide the substrate current when the NMOS is inverted, and it also provides gate leakage in the strong inversion PMOS.

The typical direct tunneling leakage characteristic in Figure 3.37 corresponds to the accumulation area of the PMOS capacitor. According to the previous model, the comprehensive current expression relationship is

$$J_{DT} = \frac{qkT}{8\pi h \phi_b} \cdot \left(\frac{V_g}{d^2} \right) \ln \left[1 + \exp \left(\frac{q(V_g - V_{FB})}{kT} \right) \right] \exp \left[\frac{20}{\phi_b} V_{ox} \left(1 - \frac{V_{ox}}{\phi_b} \right) \right]$$

$$\exp \left(- \frac{8\pi \sqrt{2m_{ox} q} \phi_b^{\frac{3}{2}} \left[1 - \left(1 - \frac{V_{ox}}{\phi_b} \right)^{\frac{3}{2}} \right]}{3hE_{ox}} \right) \tag{3.54}$$

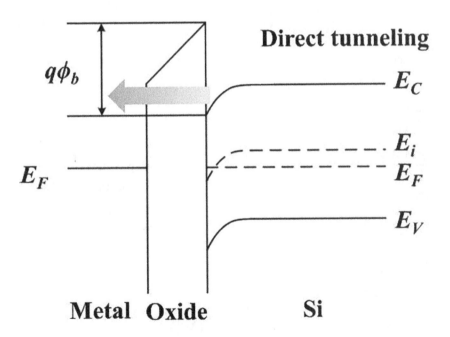

FIGURE 3.37   Schematic diagram of effective barrier height during direct tunneling.

There are two fitting parameters: one is the effective electron effective mass, m_{ox}, of the gate dielectric, and the second one is the effective barrier height, ϕ_b, of the metal to the semiconductor substrate. The physical meaning of the barrier height is shown in Figure 3.37.

Finally, the physical parameters used in the fitting are $m_{ox}=0.24m_0$, $\phi_b=2.6\,\text{eV}$, and the fitting result obtained is shown in Figure 3.38.

3.5.2 Poole–Frenkel Leakage

Some shallow-level impurities are usually introduced in high-k materials due to process reasons. These impurities will pass through the Poole–Frenkel emission process and contribute to the tunneling current of the entire gate dielectric. The Poole–Frenkel emission process means that carriers first enter the defect state in the gate dielectric through a tunneling process and then transition to the substrate through a smaller energy transition, thereby forming a current. This process is related to the energy-level position of the defect state and the material properties of the gate dielectric. The leakage current due to the Poole–Frenkel (P–F) mechanism is given as:

$$J_{PF} = A_{PF} E_{ox} \exp\left[\frac{-q(\phi_t - \sqrt{qE_{ox}/\pi\varepsilon_{ox}})}{kT} \right] \tag{3.55}$$

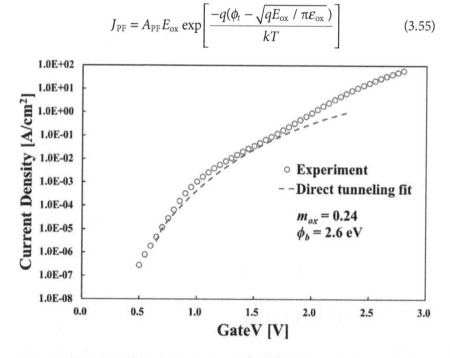

FIGURE 3.38 Fitting of direct tunneling to experimental results.

Among them, the defect emission process is related to temperature, A_{PF} is the undetermined coefficient of P–F emission, E_{ox} is the effective electric field in the oxide layer, ϕ_t is the energy-level position of the defect relative to the conduction band of the dielectric layer, and k is the Boltzmann constant.

According to the I–V characteristic curve in Figure 3.38, draw the curve of J/E_{ox} and $E_{ox}^{1/2}$ and analyze the P–F emission process according to the relationship of formula (3.12). By fitting the linear part, the relationship between the defect and the medium can be obtained. In the relationship between J/E_{ox} and $E_{ox}^{1/2}$, as shown in Figure 3.39, the linear part contains two segments. The first segment is the part with a smaller slope below 1.4 MV/cm, and the second segment is 1.7 ~ 3.2 MV/cm part. In the range of small electric field, although J/E_{ox} and $E_{ox}^{1/2}$ follow a linear relationship, the fitted dielectric constant value is greater than 100, which is far from the actual, so it cannot be explained by the P–F emission process. For the higher voltage area, the slope of the straight line can be obtained as follows:

$$q\sqrt{\frac{q}{\pi\varepsilon_{ox}}}\frac{1}{kT} = 29.707\sqrt{\mathrm{cm}/\mathrm{MV}} \qquad (3.56)$$

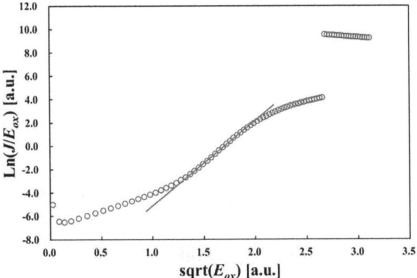

FIGURE 3.39 J/E_{ox} and $E_{ox}^{1/2}$ relationship.

According to this relationship, the dielectric constant of the dielectric layer can be calculated to be 15.9, which is in line with the theoretical value range of HfO_2/IL.

The physical process of P–F emission is shown in Figure 3.40. The carriers first enter the defect state energy level of the gate dielectric through the tunneling process and then pass a smaller energy transition to the substrate to form a current. It can be seen that the defect energy level ϕ_t is the distance between the defect energy level and the conduction band of the dielectric layer, which is much smaller than the metal-to-medium barrier height. Therefore, to study the defect-level position during the P–F emission process, it is necessary to compare the changes of the I–V curve at different temperatures. First, we normalized the relationship between I–V characteristics and temperature between 1.3 and 1.8 MV/cm, as shown in Figure 3.41, and examined the relationship between current increment and voltage at 25°C at different temperatures. It can be seen from the figure that at about 2 V, the incremental value of the current reaches a peak value, which is about 160% increase.

Second, through the Arrhenius plot of the current at different temperatures, the relationship between the effective activation energy and the electric field can be seen, as shown in Figure 3.42. From the intercept in the figure, the depth of the defect level is about 0.2 eV.

Poole-Frenkle emission

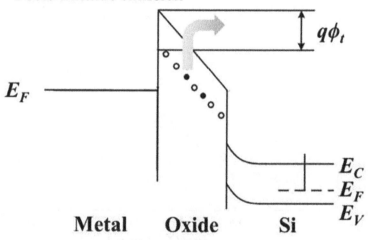

FIGURE 3.40   Poole–Frenkel emission process.

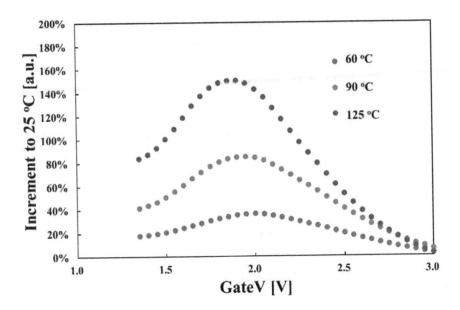

FIGURE 3.41   Normalized I–V characteristics at different temperatures.

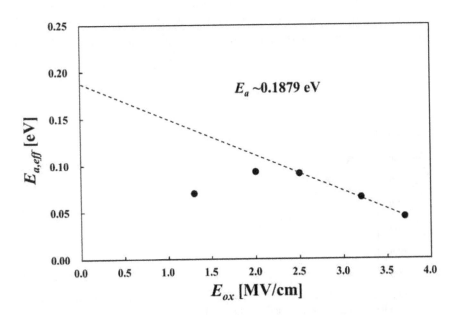

FIGURE 3.42   The relationship between effective activation energy and electric field.

3.5.3 Fowler–Nordheim Tunneling

For high-field conditions, the Fowler–Nordheim (F–N) tunneling process is the main carrier transport mechanism in the thin oxide layer. The current expression of F–N tunneling is

$$J_{FN} = A_{FN}E_{ox}^2 \exp\left[-\frac{8\pi\sqrt{2m^*q}}{3hE_{ox}}\phi_b^{\frac{3}{2}}\right] \tag{3.57}$$

Here

$$\frac{q^2}{8\pi h} = 1.53726e - 6A\,/\,V$$

$$\frac{8\pi\sqrt{2m_0q}}{3h} = 6.8228e9V^{-\frac{1}{2}}m^{-1} \tag{3.58}$$

Therefore, by drawing the relationship between $\ln(J/E_{ox}^2)$ and $1/E_{ox}$, the linear relationship of F–N tunneling can be obtained, as shown in Figure 3.43. It can be seen that when the electric field is above 2.5 MV/cm ($1/E_{ox} < 0.4$ cm/MV), the curve in the F–N plot is approximately in line

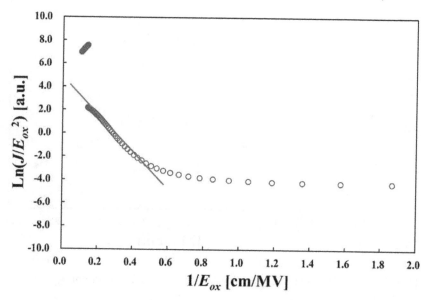

FIGURE 3.43 F–N plot relationship of I–V characteristic of MOS capacitor.

with the linear relationship, indicating that the conduction mechanism in the high field is determined by F–N tunneling. However, as the current increases, it can be seen that the current before breakdown has a slight downward trend compared with the straight line. Therefore, for the MOS capacitor in this article, at high field, on the one hand, F–N tunneling determines the leakage current, and on the other hand, the influence of the series resistance effect must be considered.

From the F–N plot, the effective electron mass, m_{ox}, and the effective barrier height, ϕ_b, can be fitted. Among them, m_{ox} adopts the result of the previous direct tunneling model, which is 0.24, and the fitted barrier height, ϕ_b, is 0.4 eV. The barrier height obtained is not consistent with the literature, which may be due to the influence of series resistance on the characteristics of F–N, or the result of the combined effect of other transport mechanisms.

3.5.4 Other Transport Mechanisms of Carriers

In addition to the carrier transport mechanism observed in the previous experiment, we also considered other possible situations and discussed them. In a lower electric field, the thermal emission process of electrons is generally considered, that is, the process of electrons in the metal gaining high enough energy to jump over the barrier. This process can be expressed by:

$$J_{SE} = A_{SE}T^2 \exp\left[\frac{-q(\phi_b - \sqrt{qE_{ox}/4\pi\varepsilon_{ox}})}{kT}\right] \tag{3.59}$$

A set of calculation results for the above relationship is shown in Figure 3.44, where the physical parameters used are $\phi_b = 1\,V$ and $\varepsilon_{ox} = 15.9$. The results show that the theoretical calculation result of the SE emission current is much smaller than the actual current, which shows that the SE emission current is negligible in the case of MOS structure thin oxygen.

For thinner high-k gate dielectrics, the defect-assisted tunneling process must also be considered. Defect-assisted tunneling refers to the existence of defect energy levels close to the band edge in the gate dielectric, and the carriers in the electrodes on both sides are easier to tunnel to this defect position. Generally speaking, through the capture and release process of one to two defect energy levels, the current can be formed. This kind of process usually occurs at a higher electric field, and the general current expression is

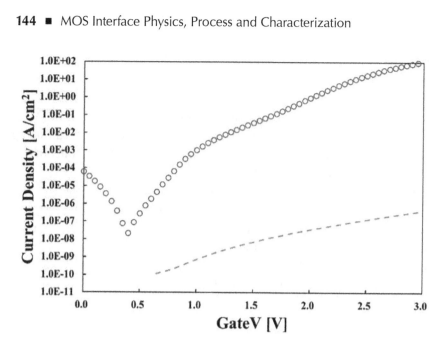

FIGURE 3.44 Comparison of SE emission calculation results with the experimental I–V curve.

$$J_{TAT} = A_{TAT} \exp\left(-\frac{8\pi\sqrt{2qm_{ox}}}{3hE_{ox}} \phi_t^{\frac{3}{2}} \right) \qquad (3.60)$$

There is no obvious linear relationship in the characteristic relationship of the MOS capacitor $\ln(J)$–$1/E_{ox}$, as shown in Figure 3.44. Therefore, for devices that have not undergone stress testing, the defect-assisted tunneling process is not significant or is annihilated under direct tunneling and F–N tunneling currents (Figure 3.45).

Based on the above analysis, without considering the multi-phonon defect-assisted tunneling process and the modified F–N tunneling process, the conclusion is that in the case of less than 1 V, the direct tunneling process is mainly considered; at $1 \sim 2$ V between is the P–F emission process; when it is higher than 2 V, it is mainly due to the combined effect of the F–N tunneling process and the series resistance effect. The overall current characteristic analysis is shown in Figure 3.46. In addition, through the fitting, we also get some physical characteristics of the oxide layer, such as the effective mass of electrons in the gate dielectric is $0.24m_0$, the relative dielectric constant is 15.9, the defect level is about

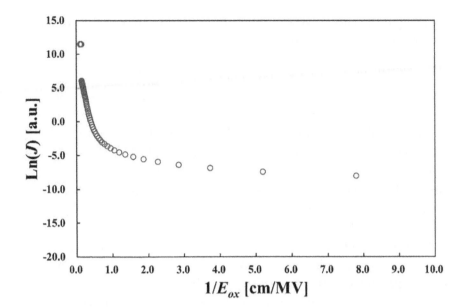

FIGURE 3.45 $\ln(J)$–$1/E_{ox}$ relationship.

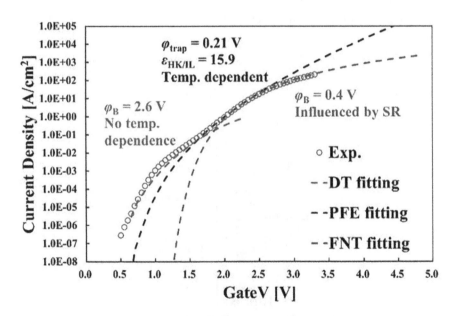

FIGURE 3.46 Schematic diagram of carrier transport mechanism corresponding to different voltage regions.

0.2 eV, and the effective potential of the metal and dielectric layers. The barrier height is 2.6 eV and so on.

3.6 WORK FUNCTION AND EFFECTIVE WORK FUNCTION

The work function is generally defined as the difference between the vacuum level and the Fermi level. After the introduction of high-k dielectric and metal gate into the modern MOS devices, the term effective work function (EWF) is more important and usually used. Thus, we discuss the EWF. Figure 3.47 schematically shows the definition of the EWF using the band alignment of the whole metal/high-k/SiO$_2$/Si stack. Before the contact of the metal, high-k dielectric, SiO$_2$ and Si, the vacuum level for the four materials is consistent.

Figure 3.48 shows the relative shift of the vacuum levels of the four materials after considering the contributions to the EWF from the following parameters: the areal charge at the high-k/SiO$_2$ interface, the bulk charges in the high-k dielectric, the dipole at the high-k/SiO$_2$ interface and the Fermi-level pinning (FLP) at the metal/high-k interface. The EWF of the metal gate is referenced to the vacuum level of the Si substrate but not the vacuum level of the metal. It should be noted that shown in Figure 3.48 is not energy band alignment at thermal equilibrium after the contact of metal, high-k, SiO$_2$ and Si substrates.

Then using the concept of EWF of the metal gate, a simplified model can be used to conveniently predict the band bend of the Si substrate

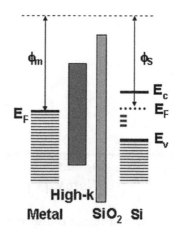

FIGURE 3.47 Schematic of energy band before contact of metal, high-k dielectric, SiO$_2$ and Si substrates.

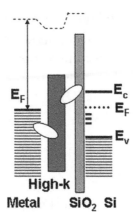

E_F E_c E_F E_v

High-k
Metal   **SiO₂ Si**

FIGURE 3.48 Schematic of energy band after contact of metal, high-k dielectric, SiO$_2$ and Si substrates.

and further guide the gate engineering of CMOS devices. This simplified model can be equivalent to the structure of metal (with vacuum work function of EWF)/SiO$_2$ (with the thickness of EOT)/Si substrates as shown in Figure 3.49. The band bend of the Si substrate of this equivalent stack is the same as that of the metal/high-k/SiO$_2$/Si stack. Thus, the V_{th} is also consistent for these two stacks. Therefore, the introduction of EWF can make the analysis of the MOS structure with high-k/metal gate stacks convenient due to the simplified conventional model of poly-Si/SiO$_2$/Si structure.

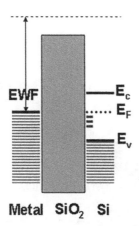

EWF E_c E_F E_v

Metal SiO₂ Si

FIGURE 3.49 Schematic of a simplified equivalent metal/SiO$_2$/Si structures in terms of EWF concept.

3.6.1 Definition of EWF Based on Terraced SiO$_2$

For MOS capacitors with metal/high-k/terraced SiO$_2$/Si stacks as shown in Figure 3.50, the V_{FB} of this structure is given as follows:

$$V_{FB} = \phi_{ms} - \frac{Q_{SiO_2,Si} \text{EOT}}{\varepsilon_0 \varepsilon_{SiO_2}} - \frac{\rho_{bulk,SiO_2} \text{EOT}^2}{2\varepsilon_0 \varepsilon_{SiO_2}} - \frac{Q_{high-k,SiO_2} d_{high-k}}{\varepsilon_0 \varepsilon_{high-k}}$$

$$- \frac{\rho_{bulk,high-k} d_{high-k}^2}{2\varepsilon_0 \varepsilon_{high-k}} + \frac{\rho_{bulk,SiO_2} \varepsilon_{SiO_2} d_{high-k}^2}{2\varepsilon_0 \varepsilon_{high-k}^2} + \Delta V_{high-k,SiO_2} + \Delta V_{metal,high-k} \quad (3.61)$$

where EOT is the equivalent oxide thickness of the whole meta/high-k/SiO$_2$/Si stack. ϕ_{ms} is the vacuum work function difference between the metal gate and the Si substrate. $Q_{SiO_2,Si}$ and Q_{high-k,SiO_2} are the areal charge densities (per unit area) at SiO$_2$/Si and high-k/SiO$_2$ interfaces, respectively. ρ_{bulk,SiO_2} and $\rho_{bulk,high-k}$ are the bulk charge densities (per unit volume) in SiO$_2$ and high-k dielectric. $\Delta V_{high-k,SiO_2}$ and $\Delta V_{metal,high-k}$ are the V_{FB} shift moments due to the possible dipole at the high-k/SiO$_2$ interface and FLP at the metal gate/high-k interface. ε_0, ε_{SiO_2} and ε_{high-k} express the vacuum permittivity, the relative permittivity of SiO$_2$ and high-k dielectric, respectively. d_{high-k} is the physical thickness of the high-k dielectric. Then the intercept (I) of V_{FB}–EOT plot can be obtained to be

$$I = \phi_{ms} - \frac{Q_{high-k,SiO_2} d_{high-k}}{\varepsilon_0 \varepsilon_{high-k}} - \frac{\rho_{bulk,high-k} d_{high-k}^2}{2\varepsilon_0 \varepsilon_{high-k}} + \frac{\rho_{bulk,SiO_2} \varepsilon_{SiO_2} d_{high-k}^2}{2\varepsilon_0 \varepsilon_{high-k}^2}$$

$$+ \Delta V_{high-k,SiO_2} + \Delta V_{metal,high-k} \quad (3.62)$$

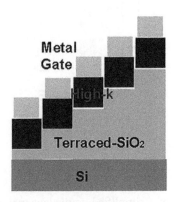

FIGURE 3.50 Schematic of metal/high-k/terraced SiO$_2$/Si structures.

Then, the EWF of the metal gate is expressed as

$$\text{EWF} = \phi_m - \frac{Q_{\text{high-k,SiO}_2} d_{\text{high-k}}}{\varepsilon_0 \varepsilon_{\text{high-k}}} - \frac{\rho_{\text{bulk,high-k}} d_{\text{high-k}}^2}{2\varepsilon_0 \varepsilon_{\text{high-k}}}$$

$$+ \frac{\rho_{\text{bulk,SiO}_2} \varepsilon_{\text{SiO}_2} d_{\text{high-k}}^2}{2\varepsilon_0 \varepsilon_{\text{high-k}}^2} + \Delta V_{\text{high-k,SiO}_2} + \Delta V_{\text{metal,high-k}} \qquad (3.63)$$

It can be concluded that the EWF of the metal gate includes not only the contribution from the vacuum work function of the metal gate, but also the contributions from the following parameters: the areal charge at the high-k/SiO$_2$ interface, the bulk charges in the high-k dielectric, the dipole at the high-k/SiO$_2$ interface and the FLP at the metal/high-k interface.

3.6.2 Definition of EWF Based on Terraced High-k Dielectric

For MOS capacitors with metal/terraced high-k/SiO$_2$/Si stacks as shown in Figure 3.51, the V_{FB} of the gate stack can be given as follows:

$$V_{\text{FB}} = \phi_{\text{ms}} - \frac{Q_{\text{SiO}_2,\text{Si}} + Q_{\text{high-k,SiO}_2}}{\varepsilon_0 \varepsilon_{\text{SiO}_2}} \text{EOT} + \frac{\varepsilon_{\text{high-k}} \rho_{\text{bulk,high-k}}}{\varepsilon_0 \varepsilon_{\text{SiO}_2}^2} d_{\text{SiO}_2} \text{EOT}$$

$$- \frac{\varepsilon_{\text{high-k}} \rho_{\text{bulk,high-k}}}{\varepsilon_0 \varepsilon_{\text{SiO}_2}^2} \text{EOT}^2 - \frac{\varepsilon_{\text{high-k}} \rho_{\text{bulk,high-k}}}{\varepsilon_0 \varepsilon_{\text{SiO}_2}^2} d_{\text{SiO}_2}^2 +$$

$$\frac{Q_{\text{high-k,SiO}_2}}{\varepsilon_0 \varepsilon_{\text{SiO}_2}} d_{\text{SiO}_2} + \Delta V_{\text{high-k,SiO}_2} + \Delta V_{\text{metal,high-k}} \qquad (3.64)$$

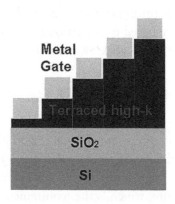

FIGURE 3.51 Schematic of metal/terraced high-k/SiO$_2$/Si structures.

Here, the d_{SiO_2} is the physical thickness of the SiO_2. The other parameters are the same as the above definitions. Then the intercept (I) of the V_{FB}–EOT plot for a metal/terraced high-k/SiO_2/Si stack can be obtained to be

$$I = \phi_{ms} - \frac{\varepsilon_{high\text{-}k}\rho_{bulk,high\text{-}k}}{\varepsilon_o\varepsilon_{SiO_2}^2}d_{SiO_2}^2 + \frac{Q_{high\text{-}k,SiO_2}}{\varepsilon_o\varepsilon_{SiO_2}}d_{SiO_2}$$

$$+\Delta V_{high\text{-}k,SiO_2} + \Delta V_{metal,high\text{-}k} \tag{3.65}$$

So the EWF of the metal gate can be expressed as follows:

$$EWF = \phi_m - \frac{\varepsilon_{high\text{-}k}\rho_{bulk,high\text{-}k}}{\varepsilon_o\varepsilon_{SiO_2}^2}d_{SiO_2}^2 + \frac{Q_{high\text{-}k,SiO_2}}{\varepsilon_o\varepsilon_{SiO_2}}d_{SiO_2}$$

$$+\Delta V_{high\text{-}k,SiO_2} + \Delta V_{metal,high\text{-}k} \tag{3.66}$$

Similarly, the EWF of the metal gate includes the contributions from the areal charge at the high-k/SiO_2 interface, the bulk charges in the high-k dielectric, the dipole at the high-k/SiO_2 interface and the FLP at the metal/high-k interface.

3.6.3 Quantitative Analysis of the Effects of Various Factors on EWF

In this section, the effects of the following parameters in the gate stack on the EWF of the metal electrode are quantitatively investigated: bulk charge density of the high-k dielectric, areal interfacial charges at the high-k/SiO_2 interface, dipole at the high-k/SiO_2 interface and FLP at the metal/high-k interface.

Figure 3.52 shows the tuning of EWF due to the interfacial charges at the high-k/SiO_2 interface based on Equation (3.61). The EWF shift due to the interfacial charges at the high-k/SiO_2 interface can be expressed as:

$$EWF = -\frac{d_{high\text{-}k}}{\varepsilon_0\varepsilon_{high\text{-}k}}Q_{high\text{-}k,SiO_2} \tag{3.67}$$

The high-k dielectrics with the permittivity of 25 and 10 are used, which contain most dielectric candidates. In addition, the density of interfacial charges is designed to be $-5\times10^{12}\,cm^{-2}$. It can be seen that when the EOT of the high-k dielectric is about 0.5 nm, which is the requirement of 45 nm technology

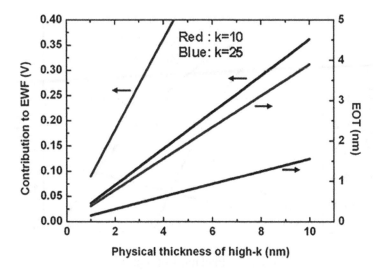

FIGURE 3.52   Effect of the interfacial charges at the high-k/SiO2 interface on the EWF of the metal gate.

node and beyond, the contribution of interfacial charges to the EWF is about 0.1 V. Compared with the dipole moment of ~0.3–0.5 V, it is negligible.

Then the contribution of bulk charges in the high-k dielectric to the EWF is clarified also based on Equation (3.61). The EWF shift due to the bulk charges in the high-k dielectric can be obtained as follows:

$$EWF = -\frac{d^2_{high\text{-}k}}{2\varepsilon_0 \varepsilon_{high\text{-}k}} \rho_{bulk,high\text{-}k} \quad (3.68)$$

In this simulation, as shown in Figure 3.53, the density of bulk charges in the high-k dielectric is taken to be $1 \times 10^{19}\,cm^{-3}$. It can be seen that when the EOT of the high-k dielectric is 0.5 nm, the contribution of bulk charges in the high-k insulator to the EWF is less than 0.05 V. This moment is much less than the magnitude of the dipole at the high-k/SiO$_2$ interface and can also be neglected. In summary, the fixed charges in the whole gate stack, including interfacial charges at the high-k/SiO$_2$ interface and bulk charges in the high-k dielectric, have negligible effects on the EWF shift of the metal electrode. Therefore, the most important factor that affects EWF shift of the metal gate is dipole formation at the high-k/SiO$_2$ interface and FLP at the metal/high-k interface.

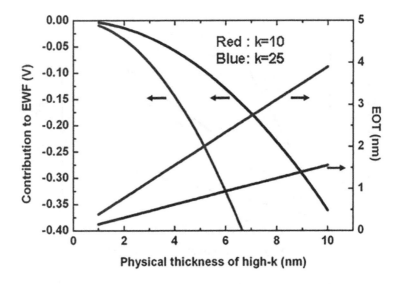

FIGURE 3.53   Effect of bulk charges in high-k dielectric on the EWF of the metal gate.

The definition of EWF of the metal gate is comprehensively studied containing various factors that affect the V_{FB} shift. The EWF controls of the metal electrode are combined results from various factors as shown in Figure 3.54.

The interface charges at the high-k/SiO$_2$ interface and bulk charges in the high-k dielectric are found to have negligible effects on the V_{FB} shift, when the EOT is scaling down below ~0.5 nm. The dipole formation at the high-k/SiO$_2$ interface and FLP at the metal/high-k interface are the dominant factors in the engineering for EWF tuning. The interfacial-induced gap states at the metal/high-k and high-k/SiO$_2$ interfaces play an important role in the FLP phenomenon at the metal/high-k interface and dipole formation at the high-k/SiO$_2$ interface. Thus, the passivation of the metal/high-k and high-k/SiO$_2$ interfaces is essential to obtain appropriate band-edge EWF for future CMOS technology beyond 32 nm. Hafnium-based oxides present huge potentials for applications in the future, especially HfO$_2$. The necessary deposition and processing to produce working devices have been achieved. However, the oxides need to be optimized substantially further, in order to achieve high-performance devices.

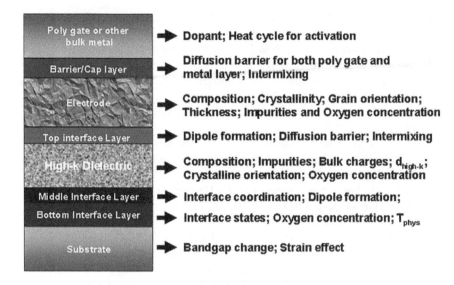

FIGURE 3.54   Summary of factors affecting EWF of the metal electrode.

BIBLIOGRAPHY

1. Terman L.M., An investigation of surface states at a silicon/silicon oxide interface employing metal-oxide-silicon diodes[J]. *Solid-State Electronics*, 1962. **5**(5): p. 285–299.
2. Berglund, C.N. Surface states at steam-grown silicon-silicon dioxide interfaces [J]. *Electron Devices IEEE Transactions*, 1966. **ED13**(10): p. 701–705.
3. Castagne, R., and A. Vapaille, Description of the SiO_2 Si interface properties by means of very low frequency MOS capacitance measurements [J]. *Surface Science*, 1971, **28**(1): p. 157–193.
4. Yoshioka, H., T. Nakamura, and T. Kimoto, Accurate evaluation of interface state density in SiC metal oxide semiconductor structures using surface potential based on depletion capacitance[J]. *Journal of Applied Physics*, 2012. **111**(1): p. 04C100.
5. Nicollian E.H., J.R. Brews, and E.H. Nicollian, *MOS (Metal Oxide Semiconductor) Physics and Technology [M]*, (Wiley, New York, 1982).
6. Yu, B., Y. Yuan, H.P. Chen, J. Ahn, P.C. McIntyre, and Y. Taur, Effect and extraction of series resistance in Al_2O_3-InGaAs MOS with bulk oxide trap[J]. *Electronics Letters*, 2013. **49**(7): p. 492–493.
7. Peng, Z.-Y., S.K. Wang, Y. Bai, Y.D. Tang, X.M. Chen, C.Z. Li, K.A. Liu, and X.Y. Liu, High temperature 1 MHz capacitance-voltage method for evaluation of border traps in 4H-SiC MOS system. *Journal of Applied Physics*, 2018. **123**: p. 135302.

8. Potbhare, S.N. Goldsman, A. Lelis, J.M. McGarrity, F.B. McLean, and D. Habersat, A physical model of high temperature 4H-SiC MOSFETs[J]. *IEEE Transactions on Electron Devices*, 2008. **55**(8): p. 2029–2040.
9. Sah, C.T., R.N. Noyce, and W. Shockley, Carrier generation and recombination in P-N junctions and P-N junction characteristics [J]. *Proceedings of the IRE*, 1957. **45**(9): p. 1228–1243.
10. Hall, R.N. Electron hole recombination in Germanium [J]. *Physical Review*, 1952. **87**(1): p. 2219.
11. Shockley, W., and W.T. Read, Statistics of the recombinations of holes and electrons [J]. *Physical Review*, 1952. **87**(5): p. 83–542.
12. Peng, Z.-Y., S.-K. Wang, Y. Bai, Y.-D. Tang, X.-M. Chen, C.-Z. Li, K.-A. Liu and X.-Y. Liu, High temperature 1 MHz capacitance-voltage method for evaluation of border traps in 4H-SiC MOS system, *Journal of Applied Physics*, 2018. **123**(13): p. 135302.
13. Wang, S.-K., B. Sun, M.-M. Cao, H.-D. Chang, Y.-Y. Su, H.-O. Li and H.-G. Liu, Modification of Al2O3/InP interfaces using sulfur and nitrogen passivations, *Journal of Applied Physics*, 2017. **121**(18): p. 184104.
14. Wang, S.-K., M. Cao, B. Sun, H. Li and H. Liu, Reducing the interface trap density in Al2O3/InP stacks by low-temperature thermal process, *Applied Physics Express* 2015. **8**(9): p. 091201.
15. Yang, X., S.-K. Wang, X. Zhang, B. Sun, W. Zhao, H.-D. Chang, Z.-H. Zeng and H. Liu, Al2O3/GeOx gate stack on germanium substrate fabricated by in situ cycling ozone oxidation method, *Applied Physics Letters*, 2014. **105**(9): p. 092101.
16. Liu, X., J. Hao, N. You, Y. Bai and S. Wang, High-pressure microwave plasma oxidation of 4H-SiC with low interface trap density, *AIP Advances*, 2019. **9**(12): p. 125150.
17. Nicollian, E.H., and J.R. Brews, *MOS Physics Technology*, (John Wiley and Sons, New York, 1982).
18. Yoshioka, H., T. Nakamura, and T. Kimoto, Accurate evaluation of interface state density in SiC metal-oxide-semiconductor structures using surface potential based on depletion capacitance, *Journal of Applied Physics*, 2012. **111**: p. 014502.
19. *Agilent Impedance Measurement Handbook: A Guide to Measurement Technology and Techniques* 4th Edition, www.agilent.com/find/impedance.
20. Sze, S.M. *Physics of Semiconductor Devices*, 3rd Edition. (Wiley, New York, 2006), Chap. 4.
21. Yuan, Y., L. Wang, B. Yu, B. Shin, J. Ahn, P.C. McIntyre, P.M. Asbeck, M.J.W. Rodwell, and Y. Taur, A distributed model for border traps in Al2O3–InGaAs MOS devices, *IEEE Electron Device Letters*, 2011. **32**: p. 485.

Appendix I: Physical Constants

Name	Symbol	Value
Absolute zero	0 K	$= -273.15°C$
Acceleration of gravity	g	$9.8 \, m/s^2$
Avogadro's number	N_0	$= 6.022 \times 10^{22}$
Coulomb constant	K	$= 8.998 \times 10^9 \, Nm^2/C^2$
Electron charge	q	$= 1.602 \times 10^{-19} C$
Gravitational constant	G	$= 6.673 \times 10^{-11} \, Nm^2/kg^2$
Mass of electron	m_e	$= 9.109 \times 10^{-31} kg$
Mass of proton	m_p	$= 1.673 \times 10^{-27} kg$
Planck's constant	h	$= 6.626 \times 10^{-34} \, Js$
Speed of light in vacuum	c	$= 2.997 \times 10^8 \, m/s$

Appendices II–V: Useful Data for MOS Interface in Periodic Table

Periodic system of the elements with work function $\varphi_{m,vac}$

1	2	3	4	5	6	7	8	9	10	11	12	13	14	15	16	17
Li 5.39	Be 4.98											B 4.45	C 5	N -	O -	F -
Na 2.75	Mg 3.66											Al 4.28	Si 4.85	P -	S -	Cl -
K 2.3	Ca 2.87	Sc 3.5	Ti 4.33	V 4.3	Cr 4.5	Mn 4.1	Fe 4.5	Co 5	Ni 5.15	Cu 4.65	Zn 4.33	Ga 4.2	Ge 5	As 3.75	Se 5.9	Br -
Rb 2.16	Sr 2.59	Y 3.1	Zr 4.05	Nb 4.3	Mo 4.6	Tc	Ru 4.71	Rh 4.98	Pd 5.12	Ag 4.26	Cd 4.22	In 4.12	Sn 4.42	Sb 4.55	Te 4.95	I -
Cs 2.14	Ba 2.7	L	Hf 3.9	Ta 4.25	W 4.55	Re 4.96	Os 4.83	Ir 5.27	Pt 5.65	Au 5.1	Hg 4.49	Tl 3.84	Pb 4.25	Bi 4.22	Po -	At -
Fr	Ra															

L	La 3.5	Ce 2.9	Pr 3.0	Nd 3.2	Pm 3.1	Sm 2.7	Eu 2.5	Gd 3.1	Tb 3	Dy 3.3	Ho 3.2	Er 3.3	Tm 3.1	Yb 3.0	Lu 3.3

Periodic system of the elements with Pauling's electronegativity

	1	2	3	4	5	6	7	8	9	10	11	12	13	14	15	16	17
2	Li 1	Be 1.5											B 2	C 2.5	N 3	O 3.5	F 4
3	Na 0.9	Mg 1.2											Al 1.5	Si 1.8	P 2.1	S 2.5	Cl 3
4	K 0.8	Ca 1.0	Sc 1.3	Ti 1.5	V 1.6	Cr 1.6	Mn 1.5	Fe 1.8	Co 1.8	Ni 1.8	Cu 1.9	Zn 1.6	Ga 1.6	Ge 1.8	As 2.0	Se 2.4	Br 2.8
5	Rb 0.8	Sr 1.0	Y 1.2	Zr 1.4	Nb 1.6	Mo 1.8	Tc 1.9	Ru 2.2	Rh 2.2	Pd 2.2	Ag 1.9	Cd 1.7	In 1.7	Sn 1.8	Sb 1.9	Te 2.1	I 2.5
6	Cs 0.7	Ba 0.9	L	Hf 1.3	Ta 1.5	W 1.7	Re 1.9	Os 2.2	Ir 2.2	Pt 2.2	Au 2.4	Hg 1.9	Tl 1.8	Pb 1.8	Bi 1.9	Po 2.0	At 2.2
7	Fr 0.7	Ra 0.9															

L	La 1.1	Ce 1.1	Pr 1.1	Nd 1.1	Pm -	Sm 1.2	Eu -	Gd 1.2	Tb -	Dy 1.2	Ho 1.2	Er 1.2	Tm 1.3	Yb -	Lu 1.3

Periodic system of the elements with melting point

	1	2	3	4	5	6	7	8	9	10	11	12	13	14	15	16	17
2	Li 179	Be 1278											B 2300	C 3500	N -	O -	F -
3	Na 98	Mg 651											Al 660	Si 1414	P -	S -	Cl -
4	K 64	Ca 848	Sc 1541	Ti 1657	V 1890	Cr 1890	Mn 1244	Fe 1535	Co 1494	Ni 1455	Cu 1084	Zn 420	Ga 30	Ge 959	As 817	Se 220	Br -
5	Rb 39	Sr 769	Y 1495	Zr 1852	Nb 2415	Mo 2610	Tc 2172	Ru 2250	Rh 1963	Pd 1554	Ag 962	Cd 321	In 157	Sn 232	Sb 631	Te 450	I -
6	Cs 29	Ba 725	L	Hf 2230	Ta 2996	W 3387	Re 3180	Os 2700	Ir 2457	Pt 1772	Au 1064	Hg -34	Tl 303	Pb 328	Bi 271	Po -	At -
7	Fr 27	Ra 700	A														

L	La 918	Ce 795	Pr 935	Nd 1024	Pm 1652	Sm 1072	Eu 1527	Gd 1312	Tb 1356	Dy 1407	Ho 1461	Er 1497	Tm 1547	Yb 824	Lu 1652

Periodic system of the elements with ionization potential

1	2	3	4	5	6	7	8	9	10	11	12	13	14	15	16	17
Li 5.4	Be 9.3											B 8.3	C 11.3	N 14.5	O 13.6	F 17.4
Na 5.1	Mg 7.6											Al 5.98	Si 8.15	P 10.5	S 10.4	Cl 13.0
K 4.34	Ca 6.11	Sc 6.54	Ti 6.82	V 6.74	Cr 6.77	Mn 7.44	Fe 7.87	Co 7.86	Ni 7.64	Cu 7.73	Zn 9.40	Ga 6.0	Ge 7.90	As 9.81	Se 9.75	Br -
Rb 4.18	Sr 5.70	Y 6.38	Zr 6.84	Nb 6.88	Mo 7.10	Tc 7.28	Ru 7.37	Rh 7.46	Pd 8.34	Ag 7.58	Cd 8.99	In 5.79	Sn 7.34	Sb 8.64	Te 9.01	I -
Cs 3.89	Ba 5.21	L	Hf 7	Ta 7.89	W 7.98	Re 7.88	Os 8.7	Ir 9.1	Pt 9	Au 9.23	Hg 10.4	Tl 6.11	Pb 7.42	Bi 7.29	Po -	At -
Fr	Ra 5.28	A														

L	La 5.58	Ce 5.47	Pr 5.42	Nd 5.49	Pm 5.55	Sm 5.63	Eu 5.67	Gd 6.14	Tb 5.85	Dy 5.93	Ho 6.02	Er 6.1	Tm 6.18	Yb 6.25	Lu 5.43